DNA BARCODING OF ETHNO-MEDICINAL SPECIES OF SOLANACEAE

LALDINFELI RALTE

TABLE OF CONTENTS

ABBREVIATIONS

NHM	National History of Museum
SGN	Sol Genomics Network
WHO	World Health Organization
TLC	Thin Layer Chromatography
HPLC	High Performance Liquid Chromatography
RAPD	Random Amplified Polymorphic DNA
ISSR	Inter-Simple Sequence Repeats
SSR	Simple Sequence Repeats
RFLP	Restriction Fragment Length Polymorphism
AFLP	Amplified Fragment Length Polymorphism
rbcL	Ribulose-bisphosphate carboxylase
matK	Maturase K
ITS2	Internal transcribed spacer gene 2
NE	Northeastern
ROS	Reactive Oxygen Species
FTIR	Fourier Transformed Infrared Spectroscopy
EP	European Pharmacopeia
QTL	Quantitative Trait Loci
T-DNA	Transfer DNA
TMV	Tobacco Mosaic Virus

GPS	Global Positioning System
ISE	International Society of Ethnobiology
CTAB	Cetyl Trimethyl Ammonium Bromide
mM	millimolar
NaCl	Sodium Chloride
EDTA	Ethylenediamine tetraacetic acid
PVP	Polyvinylpyrrolidone
µl	microliter
PCR	Polymerase Chain Reaction
BLAST	Basic Alignment Search Tool
MEGA	Molecular Evolutionary Genetics
NJ	Neighbor-Joining
g	gram
ml	milliliter
µg/ml	microgram per milliliter
mg	milligram
v/v	volume by volume

Chapter 1

Introduction

1.1. A brief history of Solanaceae

1.1.1. Solanaceae in world

Solanaceae family commonly known as nightshades belongs to the order Solanales, in the group of Asterids and class of Magnoliopsida (dicotyledons). The family is angiospermic i.e., flowering plants, and has highly diversified plants ranging from annual to perennial herbs, vines, shrubs, epiphytes, and trees (Knapp et al. 2004). Solanaceae consists of about 98 genera and 2,700 species worldwide (Olmstead and Bohs 2007) that includes several agricultural, horticultural, medicinal plants, spices, and weeds with highly diverse habitats and ecology. The family is cosmopolitan in distribution having found all over the world, including tropical temperate and desert regions except Antarctica, and is greatly abundant in South and Central America (Ganaie et al. 2018). Based on the fossil tomatillos discovered in the Patagonian region, Argentina, scientists reported that the development of Solanaceae started during the Mesozoic Era (Wilf et al. 2017). The family, Solanaceae is believed to have originated from the Andean/Amazonian region of South America (NHM 2008). The occurrence is thought to have evolved in a wide range of environments, from rain forests with over three meters of annual rainfall to desert with almost no rainfall and mountain with snow and sub-freezing temperatures (SGN 2008).

Wettstein believed that the Solanaceae had a polyphyletic origin because of their close kinship with several other plant groups (Ara et al. 2011). The family was closely related to Scrophulariaceae and was divided into the tribes: Nicandreae, Solanaceae, Datureae, Cestreae, and Salpiglossideae by Wettstein (1981). Bentham and Hooker included the family within Polemoniales, however, Hallier considered it as Tubiflorae, which had been derived from Linaceae (Ara et al. 2011). Hutchinson then combined Solanales with Convolvulaceae, which he believed was the ancestor of Solanaceae. Several related genera or groups of genera were included in this family by some researchers; others placed them in related families, and yet others divided them into other families (Ganaie et al. 2018). This inconsistency in systematic treatment is partly attributed to the prevalence of unexpected feature combinations in these linked taxa. Some may lack the characteristics of the family or have characteristics of another family (Ganaie et al. 2018). This is because numerous genera within the family have an inconsistent systematic position. The taxonomic position of the above-mentioned taxa is unknown mainly due to circumstances such as the origin of the family in a region that is yet mostly unexplored (Ganaie et al. 2018). Solanaceae, in contrast to other angiosperm families, have annual habits and they are strongly zygomorphic in their basal clades. This is likewise not the case in other angiosperms groups (Stebbins 1974). Solanaceae is an intriguing family of plants because of the large number of variations within the genus Solanum itself (Knapp et al. 2004). For these reasons, Solanaceae is regarded as the 'paradoxical family' (Knapp et al. 2004). Modern and new molecular markers, which have changed the entire landscape of biological sciences since their discovery and, application, have been employed to solve the taxonomy difficulties within Solanaceae (Ganaie et al. 2018).

Solanaceae is also one of the most economically important families among the angiosperms and contains crops such as *Solanum tuberosum* (potato), *Solanum lycopersicon* (tomato), *Solanum melongena* (eggplant), *Solanum betaceum* (tomarillo/tree tomato) *Physalis peruviana* (Cape gooseberry), *Capsicum spp.* (pepper) etc. that are important food sources in human civilizations. The family also produces important alkaloids and adds significantly to their medicinal values. Plants such as *Nicotiana spp., Atropa belladonna, Hyoscyamus, Mandragora,* and *Datura spp.*, etc. are important for the production of drugs in pharmaceutical industries while some have ornamental values (such as *Petunia, Brugmansia, Cestrum, Solanum pseudocapsicum etc.*) in horticultural fields.

1.1.2. Solanaceae in India

In India, the Solanaceae family contains 29 genera and 116 species (Kalidas and Panda 2019). Among them, 12 genera and 39 species are found in Eastern Ghats of India (Venkatappa 2011), and 8 genera, 30 species were reported to occur in West Bengal (Basak et al. 2017). The genus *Solanum* is the most abundant with 39 species distributed throughout the country (Venkatappa 2011).

In India, Solanaceae plays a significant role, which includes mostly the vegetables that are consumed every day as human diet. Eggplant, potato, pepper, and tomato are some of the most economically important food crops of India (Kalidass and Panda 2019). Besides the agricultural importance, the family also plays a significant role in medicinal and horticultural fields. *S. torvum, S. americanum, S. viarum, S. anguivi, Atropa belladonna, N. tabacum,* etc. contain important alkaloids and have significant medicinal importance in drugs discovery. While *S. pseudocapsicum, S. wendlandii, Cestrum nocturnum, Petunia spp.*, etc. are normally used as garden ornamental plants.

Solanaceae is one of the largest and most taxonomically challenging families among the angiosperm family (Frodin 2004). Worldwide, Solanaceae has tremendous values and has been studied extensively at the individual species level. However, in India detailed research even at the family level is limited. Again, identification is challenging due to high degrees of morphological heterogeneity among and within species and infra-specific groups. Some exotic and gregarious weed species found in waste areas are described as 'uninteresting weeds' discouraging plant taxonomists (Kalidass and Panda 2019). Apart from a brief reference in some regional floras, there is no comprehensive taxonomic account of this intriguing and useful family in Mizoram.

1.1.3. Morphological characteristics

Members of the Solanaceae family differ greatly in terms of morphological characters, ecology, and habit. The leaves of Solanaceae plants vary in shape and size, usually simple, sometimes highly lobed, generally alternate, petiolate, sub-sessile, and rarely sessile. The leaves may be herbaceous, hairy, leathery, or spiny with reticulated venation with no basal meristem. Inflorescence can be cymose, axillary and may reduce to a single flower. Flowers are bisexual, have radial symmetry, and have differentiated perianth with calyx and corolla usually have five petals and five sepals. The calyx is united at the base, sometimes inflated in fruit. Corolla varies greatly in shape and size typically rotate, tubular, campanulate, and funnel-shaped. Stamens are epipetalous and alternate with corolla lobes. Gynoecium is a single pistil with 2 locules and numerous ovules. Fruit can be berry, dehiscent capsule, and drupe and have axial placentation. Capsules are generally septicidal or valvate. Seeds are numerous, round or flattened, endospermic, oily without hairs. An embryo may be straight or curved with dicotyledons.

1.2. Ethnobotanical studies

Plants and animals have had an intrinsic interaction since ancient times and have been the factors influencing human civilization, particularly in medicinal fields (Yeung et al. 2020). Ethnobotany is the scientific study of practical uses of plants and their indigenous or traditional knowledge of local people (Iwu 2002). Various wild and cultivated plants play a significant role in human beings particularly among the tribal communities and hence they are included in their culture, customs, traditional healing against diseases, and rituals (Sajem et al. 2008). Traditional remedies are sometimes the only available therapeutics in some places. Most of the people living in rural and remote areas mainly rely on traditional remedies to cater their primary health care needs. According to World Health Organization (WHO), around 80% of the world's population mainly relies on traditional herbal remedies for their healthcare system (Inglis 1994). Nowadays, herbal products are gaining interest and their demands are increasing significantly. However, overexploitation of medicinal plants may lead to the degradation of resources, loss of biodiversity, and result in declining of these plant species over the years (Kala and Sajwan 2007).

The Mizo people harbor significant knowledge on the traditional use of medicinal plants (Ralte et al. 2021). The Mizo tribe is mainly forest dwellers that rely on shifting cultivation for their livelihood and the majority of the population live in rural areas and most of their resources such as timber, food, medicinal, etc. are obtained from the forest and hence they have a plethora of traditional knowledge on the uses of different plant products (Ralte et al. 2021). Many ethnobotanical studies around the world, including Mizoram, attempt to record the use of herbal plants for healing, which has been practicing in their different civilizations for multiple generations. Although the cultural diversity of Mizoram contributes to the richness of its traditional knowledge, access to it is limited (Ralte et al. 2018). Traditional knowledge is typically transmitted verbally and is generally person-specific (Sabran et al. 2016). As a result, only tribal leaders, village chiefs, or traditional healers in a given group or tribe frequently own the expertise (Supiandi et al. 2019). Traditional medicinal knowledge is imperiled due to the lack of written records and restricted succession patterns. As a result, research and documentation of therapeutic plants, as well as the associated traditional wisdom are required.

1.3. Molecular Studies

Traditionally, plant species identification is mainly based on morphological traits (Heinrich 2007). However, similar morphological traits are very common among the species of the Solanaceae family, especially the genus *Physalis* and *Solanum* (Feng et al. 2018). Their floral patterns are highly similar which makes the identification very difficult. Therefore, identification based on morphological characters is sometimes not reliable and efficient (Ali et al. 2014). Chemical authentication focuses on the examination of chemical constituents and differentiation is accomplished by the use of characteristic compositions (Zhang et al. 2007). Thin-layer chromatography (TLC) and high-performance liquid chromatography (HPLC) are widely used methods for determining the chemical constituents of medicinal plants. For example, TLC was employed to identify *Tribulus terrestris* (Zhang et al. 2000) and *Fructus xanthii* (Yin et al. 2005), HPLC had been used to investigate the chemical profiles of *Cassia* bark (He et al. 2005), create fingerprints for *Psoralea corylifolia* (Zhao et al. 2005), authenticate *Ephedra* (Schaneberg et al. 2003) and assess the quality of *Salvia miltiorrhiza* (Hu et al. 2005). But, because of their diverse sources and chemical complexity, the use of chromatographic methods and marker compounds to standardize botanical preparation has limitations (Joshi et al. 2004). Variability in the flavors, aromas, and physical properties of wine and coffee from year to year and area to region serves as an example (Joshi et al. 2004). Growing circumstances, harvesting time, post-harvest operations, and storage can also influence a species' chemical makeup and relative abundance (Zhang et al. 2007). The change of chemical constituents may impede verification, and in rare cases, this may be misleading if the samples are intentionally adulterated with a marker component (Zhang et al. 2007). Furthermore, due to similar chemical compositions, it may be difficult to recognize closely related species (Joshi et al. 2004; Zhang et al. 2007). Species identification is important for monitoring a wide range of biodiversity and conservation (Desalle and Amato 2004). Various types of molecular techniques such as Random Amplified Polymorphic DNA (RAPD), Inter-Simple Sequence Repeats (ISSR), Simple Sequence Repeats (SSR), Restriction Fragment Length Polymorphism (RFLP), and Amplified Fragment Length Polymorphism (AFLP) have been employed in many taxonomic and phylogenetic studies of plants (Ganaie et al. 2018). However, due to the species specificity, which required correct DNA sequences of the species, their applicability is sometimes limited to a single species (Galimberti et al. 2014). Therefore, a reliable and efficient technique of identification is needed. DNA barcoding is an ideal candidate for species identification. DNA barcoding is a technique that uses a short, standard DNA region which is universally occurred in target sites and has enough sequence variation for species discrimination (Hebert et al. 2003; Savolainen et al. 2005). DNA barcoding aims to authenticate species by matching their sequences to the assembled reference libraries of barcode sequences and to facilitate the discovery of novel species (Hebert et al. 2003; Stoeckle 2003). DNA barcodes have been widely used nowadays, which includes phylogenetic analysis, genetic diversity (Wattoo et al. 2016), ancestral inheritance (Son et al. 2003), flagging of new species (Hebert et al. 2004), food traceability (Galimberti et al. 2014), biodiversity (Yessoufou et al. 2013) and conservation assessment (Levin et al. 2003; Ashfaq et al. 2013), plant invasion ecology (Daru et al. 2016) and confirmation and authentication of medicinal plants (Xue and Li 2011), etc. In plants, *rbcL, matK, psbA-trnH, rpoC1,* and *ITS2* regions been commonly used as DNA barcodes (Liu et al. 2012). Among them, *ITS2* and *rbcL* regions have been shown to be more effective and applicable in species discrimination among families like Asteraceae, Rutaceae, Rosaceae, Araliaceae (Hollingsworth et al. 2009; Gao et al. 2010; Luo et al. 2010;

Yao et al. 2010; Crautlein et al. 2011; Pang et al. 2011; Liu et al. 2012). Hence, the present study was designed to discriminate species belonging to Solanaceae from Mizoram, India using DNA barcodes. Studies on Solanaceae plant species have not been done in Mizoram particularly on the identification of species using DNA barcodes. The present study aims at an accurate identification of Solanaceae species using DNA barcodes that will be useful for future conservations and utilization programs. The study also attempts to show the accuracy and applicability of *rbcL* and *ITS2* as barcodes for the identification of Solanaceae species found in Mizoram, India.

1.4. Phytochemical studies

Plants have been valuable sources of medicinal agents and are widely used for therapeutic purposes since time immemorial. They not only provide essential nutrients for humans but also provide biologically active compounds that are beneficial for human health and the treatment of various diseases (Liu 2003). Nowadays, medicinal plants have been gaining special interest since they can provide numerous benefits to mankind and society in medicinal and pharmacological fields as well as food sources. Hence, the scientific community's interest in studying plant-derived bioactive compounds is growing worldwide, particularly in developing nations where herbal remedies are frequently used to meet basic health needs (Yadav 2018). In India, especially in the rural northeastern (NE) region, the use of medicinal plants for their healthcare system is very common due to the availability of vegetation and socio-economic conditions of the people from the region (Ningombam and Singh 2014). Plants have a wide range of bioactive compounds that have antimicrobial, anticancer, anti-inflammatory, and antioxidant potential. Plant-derived medicines are frequently prepared from crude extracts that include a complex combination of various phytochemicals and are utilized to treat both chronic and infectious diseases (Sahoo and Manchikant 2013). Although various plants contain a vast pool of bioactive compounds, only a few have been identified and confirmed to be significant sources of bioactive compounds. Phytochemical is a natural compound found in plants that can act as a defense mechanism against diseases or to protect against diseases (Krishnaiah et al. 2009). Medicinal plants contain organic compounds such as tannins, alkaloids, carbohydrates, terpenoids, steroids, and flavonoids (Mann 1978) that are synthesized by the secondary metabolism of living organisms. Secondary metabolites are compounds with different functions and are widely used in the treatment of human, veterinary, pharmaceutical, agriculture, and scientific research (Vasu et al. 2009). These compounds are mainly derived from barks, flowers, leaves, roots, fruits, and seeds (Cowan 1999). High activity profile drugs have already been developed from biologically active compounds from medicinal plants. Interestingly, crude extracts from medicinal plants are reported to be more biologically active than isolated compounds because of their synergistic effect (Jana and Sekhawat 2010). Solanaceae is one of the largest families of angiosperms and species of this family are rich in secondary metabolites and hence play significant roles in economic, agricultural, and pharmaceutical aspects (Oliveira et al. 2006; Ralte et al. 2021). In my present study, qualitative and quantitative analyses were carried out in ten selected plants of Solanaceae species from Mizoram.

1.5. Antioxidant studies

Studies on bioactive compounds with antioxidant properties showed extensive growth, as considerable amount of evidence have revealed that oxidative stress is a key factor in aging and the development of various pathologies including autoimmune diseases, infectious, inflammatory, and cancer diseases (Ulewicz-Magulska and Wesolowski 2019; Souza et al. 2020). Therefore, the necessity of finding new bioactive compounds with high antioxidant properties is highlighted, since antioxidants can protect, stabilize, and inhibit free radicals from attacking a biological target in cells (Souza et al. 2020). Plants are nature's greatest gift to humans and are a potential source of natural antioxidants. Medicinal plants also have been gaining attention due to the presence of natural antioxidants and their association with health benefits (Arnous et al. 2001). Then, with the rising safety issues associated with the consumption of synthetic antioxidants, researchers are looking for cheaper and lesser side effect antioxidants from natural sources, particularly plants. Medicinal plants produce various antioxidative compounds to prevent reactive oxygen species (ROS) (Lu and Foo 1995). ROS are products of normal cellular metabolism which at high concentrations may cause damages to cellular structure like, nucleic acids, proteins, and lipid (Valko et al. 2007), while uncontrolled substrate can lead to the development of diseases like cancer, hypertension, diabetes, cardiovascular and neurodegenerative disorder (Mayne 2003). Polyphenols are the main plant bioactive compounds characterizing antioxidant potential due to their redox properties (Zheng and Wang 2001). They can absorb and neutralize free radicals, quenching singlets, and triplet oxygen and decompose peroxides. Flavonoids are the most frequently and widely distributed group of phenolic compounds in plants and are present in most plants, therefore, they are considered to prevent free radical associated damages by various ways such as scavenging free radicals and enzymes inhibition which is involved in the production of free radicals (Sandhar et al. 2011). Therefore, studies on discovering natural antioxidants from plants have been on the rise and various extracts and isolated compounds have been studied for their antioxidant activities using different techniques (Boligon et al. 2009; Dudonne et al. 2009; Moon and Shibamoto 2009).

1.6. Nutritional Studies

In developing countries, wild edible plants provide nutrition, food security, and income generations (Omoti and Okyi 1987; Antia et al. 2006; Dhellot et al. 2006a) and also serve as crucial constituents of human diet providing minerals, vitamins in addition to proteins and carbohydrates (Fleuret 1979; Onyenuga and Fetuga 1995; Edmonds and Chweya 1997). Vegetables are an important part of the human diet and play a crucial role in nutrition, particularly as sources of phyto-nutraceuticals (vitamins, minerals, dietary fibers). Solanaceae has one of the largest edible sources of plants. Many members of the family contain powerful alkaloids that are extremely valuable when it comes to nutritional value (Kumari et al. 2017). However, these edible plants are yet to be adequately studied and utilized. Mizoram is a hilly state in the northeastern region of India and the majority of the region is botanically unexplored or undiscovered. In Mizoram, the forests provide a large number of plants whose various parts like fruits, seeds, tubers, shoots make a significant contribution to the dietary source of the tribal people. The present study attempts to explore the nutritional status of selected Solanaceae plants species reportedly consumed by the local people of Mizoram.

1.7. Antimicrobial studies

Infectious diseases, especially in developing nations are a major source of morbidity and mortality. Antibacterial drugs of both synthetic and semi-synthetic origins are widely accessible today for the control of microorganisms (Heritange et al. 1999). However, bacterial resistance to antibacterial treatments is also fast increasing (Andersson and Hughes 2010). Globally, antibiotic resistance has become a major concern due to the non-selective use of commercial antimicrobial drugs commonly occurred in the treatment of various infectious diseases (Westh et al. 2004). Then, antibiotics also induce a variety of adverse reactions, including hypersensitivity and immunosuppression, in addition to the beneficial effects of bacterial control (Panigrahy et al. 1979). Due to these side effects, as well as the ongoing development of bacterial resistance, there is a constant need to create better antimicrobial medicines that are effective against microbial while also being less hazardous to the human. As a result, the pharmaceutical sector has driven to create new antimicrobial medications. Plants containing bioactive compounds have been successfully used all over the world to combat infectious diseases and are the main source of potential and powerful drugs (Sivastava et al. 1996). Plants' potential as a source of novel pharmaceuticals is still largely untapped and only a small fraction of the estimated 250,000-500,000 plant species have been subjected to screening of biological or pharmacological potential (Stankovic et al. 2016). Medicinal plants that are used traditionally produce various compounds having therapeutic properties (Harborne and Baxter 1995). Those compounds or substances that can inhibit or kill pathogens can be ideal candidates for the development of antimicrobial drugs. So, antimicrobial properties of medicinal plants have gained interest all over the world (Saxena and Sharma 1999). Traditionally used medicinal plants that produce herbal remedies can be important sources for producing new antibiotics (Okpekon et al. 2004) and some compounds have already been produced that are effective against antibiotic-resistant bacterial strains (Kone et al. 2004). Therefore, further research on traditionally used medicinal plants is an important task (Romero et al. 2005) and it will lead to the synthesis of more effective drugs with less toxicity (Manna and Abalaka 2000). In my present work, selected

Solanaceae plants with promising biological activity were screened and evaluated for their antibacterial potentials.

1.8. Fourier Transformed Infrared Spectroscopy (FTIR) analysis

Plants are essential bio-resources for medications in the traditional health care system. Today, people are increasingly focusing on healthy lifestyles and the use of herbal products. It is understood that the bioactive compounds present in medicinal plants, not only aid in the treatment of diseases, but can also be used as a preventative measure to improve human health (Brangule et al. 2020). The plant chemical contents differ depending on species, geographical distribution, age, temperature, and other factors (Heinrich 2015). Consequently, extensive research into effective analytical methods is required to undertake easy, reliable, and speedy control of herbal medicine production (Bostijin et al. 2018). WHO and European Pharmacopeia (EP) have prepared a guideline for assessment of the quality of herbal products and identification of functional groups using chromatography techniques (Yang et al. 2013; Kitanov et al. 2015). However, various critical disadvantages such as complicated procedures, difficulty in sample preparation, and time-consuming had been the common issues in chromatography techniques (Peerapattana et al. 2015). FTIR is one of the widely used methods to identify chemical compounds and has been used to analyze medicine for pharmacopeia in different countries (Subashini et al. 2015). FTIR is a technique that is based on the vibration of high polar bonds and functional groups of the analyzed components (Kim et al. 2004). This technique uses a minimal amount, is quick, easy, and accurate (Struat 2004), and does not require a reagent, making it more environmentally friendly (Lin et al. 2011), as a result, it is a powerful analytical tool. FTIR analysis provides a significant method for herbal analysis (Cheng et al. 2010; Bunaciu et al. 2011) and quantitative analysis of drugs (Bunaciu et al. 2010) and has also been used for discriminating related microbial strains (Wenning et al. 2002). My present study for the first time attempts to identify the functional groups present in Solanaceae plants species of Mizoram using FTIR.

Chapter 2

Review of Literature

Since time immemorial, the Solanaceae family has been playing significant roles in food, ornaments, medicines, and aesthetic plants. Over the last century, some of the species from Solanaceae have been subjected to traditional and molecular genetic researches. The first known reference of Solanaceae is in Dioscorides Codex (AD 815), in which few species from the genera and *Physalis* have been mentioned for their pharmacological significance (D'Arcy 1979). While Casper Bauhin's (1623) concept of Solanaceae as a group was gaining more acceptance. Then, Linnaeus (1753) followed Casper's work and proposed two groups. Laurent de Jussieu (1789) established the Solanaceae family, which had since been reviewed and re-examined by various taxonomists. The family is non-indigenous to the Indian subcontinent and limited work has been done on the taxonomy. The first record of this family dates back to 1100 A.D. by Charaka and Susrute (Deb 1979). Roxburgh (1832) described 27 species of Solanaceae in Flora Indica. Clarke (1883) worked on Solanaceae for Hooker's The Flora of British India and included 52 species. After the establishment of the Botanical Survey of India in 1954, a large number of plant materials have been gathered and stored in the Indian herbaria, and multiple regional and state floras of India have been published (Deb and Dutta 1974; Chowdhery and Wadhwa 1984). However, the family Solanaceae has been neglected since a comprehensive account of taxonomy was lacking (Deb 1980).

Meanwhile, Hugo de Vries and Karl Correns rediscovered and confirmed Gregor Mendel's pioneering work at the turn of the twentieth century where species of some genera of the Solanaceae became subjects of genetic research as model plants. The cultivated tomato and its wild counterparts (*Lycopersicon*), tobacco (*Nicotiana*), and petunia (*Petunia*) were predominantly used as model plants in classical and then cellular and molecular genetics. Crop plants of global importance included cultivated potatoes, eggplant (*Solanum*), and pepper species (*Capsicum*) (Ralte et al. 2021).

Except for potato tuber growth, these species are less appropriate as models for fundamental research. Classical and molecular genetic research on potato, eggplant, and pepper was thus primarily targeted at solving agricultural challenges, such as disease resistance and the development of improved cultivars. In a first conclusive report on the regeneration of haploid embryos from pollen grains, anthers of *Datura innoxia* were cultured *in vitro* (Guha and Maheshwari 1964). This was the start of the production of double haploid and as a result, a homozygous plant that led to significant application in biotechnology and plant breeding studies of self-fertile crop plants and demonstrated that homozygous plants could be obtained in a single step in a much shorter time in comparison to numerous generations of selfing. Biochemical mutant cell lines were then obtained by complete selection from leaf mesophyll protoplasts of anther culture-derived haploid *Hyoscyamus muticus* plants (Guha and Maheshwari 1964) that was similar to the first biochemical mutant selected in *Neurospora crassa* (Beadle and Tatum 1941). In *Physalis* species, the molecular foundation and possible mechanism of the inflated-calyx syndrome, a unique development of the sepals after fertilization in various Solanaceae genera were investigated (He and Saedler 2005). Among the Solanaceae species, tomato (*Lycopersicon*), tobacco (*Nicotiana*), petunia (*Petunia*), Pepper (*Capsicum*) are the most studied species. The discovery and analysis of gene function (Menda et al. 2004), linkage mapping based on morphological and isozyme using restriction fragment length polymorphisms (RFLP) (Bernatzky and Tanksley 1986; Helentjaris et al. 1985), DNA variation and construction in intraspecific populations based on single nucleotide polymorphism (SNP) markers (Shirasawa et al. 2010), genome sequencing (Bolger et al. 2014), gene cloning (Martin et al. 1993a), QTL mapping (Grandillo et al. 2013), transcriptional regulators in fruit ripening (Giovannoni 2007), FLAVR SAVR™ tomato with a prolonged shelf life for a commercialized transgenic crop (Kramer and Redenbaugh 1994) have been studied for *Lycopersicon*. The in vitro multiplication of plant cells and tissue culture (Sussex 2008), microbial culture for isolation of biochemical mutants (Negrutiu et al. 1984), the transformation of tobacco cells with T-DNA or Ti-plasmid carrying *Agrobacterium tumefaciens* (Chilton et al. 1977), self-incompatibility (Anderson et al. 1986), host-pathogen system – tobacco mosaic virus (TMV) (Scholthof 2008), plant interactions with insect herbivores (Schuman and Baldwin 2016), linkage maps with microsatellite markers (Bindler et al. 2007), genome sequences (Bombarely et al. 2012) of *Nicotiana* have been studied. The resistance to potato viruses (Cockerham 1970), hybrid bleeding (Lindhout et al. 2011), linkage maps construction using RFLP markers (Bonierbale et al. 1988), intraspecific and interspecific crosses (Gebhardt et al. 1989, 1991), whole-genome sequencing (Doganlar et al. 2002a; Hirakawa et al. 2014) have been studied on potato (*S. tuberosum*). The biosynthesis of the alkaloid (Stewart et al. 2005), cytoplasmic inheritance (Ikeno 1917), linkage map using RFLP (Prince et al. 1993), male sterility (Ramchiary et al., 2014), plant resistance gene (Romer et al. 2007), DNA-based marker for breeding applications (Holdsworth and Mazourek 2015; Ramchiary et al. 2014), genome sequencing (Kim et al. 2014) have been done on pepper (*Capsicum*). The RFLP based linkage map construction (Doganlar et al. 2002a), the QTL mapping of morphological and biochemical fruit characters (Doganlar et al. 2002b; Portis et al. 2014), the chlorogenic acid biosynthetic pathway (Gramazio et al. 2014), genome sequencing (Hirakawa et al. 2014) have been studied on eggplant (*S. melongena*).

The technique of DNA barcoding was introduced in 2003 for accurate identification of species and was designed for animals (Hebert et al. 2003, 2004b). On the other hand, a standard DNA barcode for plants was not immediately successful or accepted in the botanical world until several years later (Kress 2011). However, after the successful uses of gene regions from the mitochondrial, plastids, and nuclear genomes (Chase et al. 2005; Kress et al. 2005; Kress & Erickson 2007; Lahaye et al. 2008; Newmaster et al. 2008) four gene regions i.e., *rbcL, matK, trnH-psbA* and *ITS* have been accepted as the standard DNA barcodes for plants (CBOL Plant Working Group 2009; Li et al. 2015). Till today, the DNA barcoding technique has been popularly used for various applications and is gaining more interest in the taxonomic field also (Chen et al. 2014; Kress 2017). Costion et al. (2011) used three gene regions (*rbcL, matK,* and *trnH-psbA*) for DNA barcodes to estimate species diversity in tropical rain forest plots in Queensland, Australia. Muellner et al. (2011) used DNA barcodes for the identification of commercially important tree species of the family, Meliaceae, and concluded that *ITS* was able to identify species of this family. Nithaniyal et al. (2014) used DNA barcodes to accurately identify wood samples collected at timber processing in Andhra Pradesh and Tamil Nadu in India. Similarly, Bolson et al. (2015) also identified numerous threatened commercial trees species of the family Lauraceae. For plants DNA barcodes, *rbcL, matK, psba-trnH, rpoC1* and *ITS2* have been widely used. Chen et al. (2010) showed that *ITS2* could be a universal barcode for the identification of plants species due to the high accuracy identification rate (92%) of over 6600 samples in the seven studied phyla (Angiosperms, Gymnosperms, Ferns, Mosses, Liverworts, Algae and Fungi). On the other hand, *rbcL* gene also has several advantages such as higher success rate of PCR amplification, sequence quality, and universality of standard primers and hence considered as the best-characterized gene sequence (CBOL Plant Working Plant 2009; Vijayan and Tsou 2010). DNA barcoding is applicable in species discrimination among plants within families such as *Asteraceae, Rutaceae, Rosaceae, Areliaceae, Zingiberaceae* (Gao et al. 2010; Luo et al. 2010; Yao et al. 2010; Pang et al. 2011; Liu et al. 2012; Chen et al. 2015). Ngan et al. (1999) used DNA barcodes for the identification of ginseng. Among the family of Solanaceae, DNA barcoding of the genus *Physalis* (Feng et al. 2017), *Datura* (Bye and Sosa 2013), *Lycopersicon* (Caprar et al. 2017), *Solanum nigrum* (Wattoo et al. 2016), *Lycium barbarum* (Xin et al. 2003) had been done recently.

The family is also rich in economically significant species and plays an important role in nutritional and pharmacological industries. Species of the family have been widely used in traditional medicine for preventive and curative purposes. They are also known for producing secondary metabolites such as phenol, alkaloids, flavonoids, tannins, etc. that are used in the pharmaceuticals industries for their medicinal properties such as antioxidant, antibacterial, hepatoprotective, antipyretic, and anti-inflammatory properties (Vijaya et al. 2013; Kumar and Pandey 2014). Certain pharmacological and nutritional studies have been carried out to identify and validate the traditional medicinal applications of some plants of Solanaceae and the studied species included *Solanum torvum* (Kalita et al. 2017), *S. nigrum* L, and *S. myriacanthus* Dunal (Gogoi and Islam 2012), *S. macrocarpon* Linn (Dougnon et al. 2012), *S. schimperianum, S. nigrum, Physalis lagascae, Withania somnifera* (Almoulah 2017), *S. agrarium, S. lycocarpum, S. palinacanthum, S. paniculatum, S.*

stipulaceum (Matias et al. 2019), *P. minima L* (Saripalli et al. 2013), *Capsicum frutescens L.* (Vinayaka et al. 2010).

Chapter 3

Methodology

3.1. Study area

Mizoram is a landlocked state which lies in the northeastern region of India, with an area covers 21,087 sq. km with 21°56'N to 24°31'N latitude and 92°16'E to 93°26'E longitude. The southern part shares an international border with Myanmar and Bangladesh while the northern part shares its boundary with the states of Manipur, Assam, and Tripura. The climate of Mizoram is relatively cool in summer while the winter temperature ranges from 7 to 22°C. The average rainfall of the state is 254 centimeters per annum. The forest classification of Mizoram is Tropical Wet Evergreen Forest, Montane sub-tropical forest, Temperate Forest, bamboo forest, Quercus Forest, and Jhum-land. Mizoram has the third highest forest cover (39,40,000 acres) and the highest percentage area (90.68%) of forests covered among the states of India. The most common vegetation found in Mizoram is tropical semi-evergreen, tropical moist deciduous, sub-tropical forest. Agriculture has traditionally been a subsistence profession in Mizoram where more than 70% of its population is engaged in agriculture. Jhum cultivation has been a historic tradition of Mizoram acquired by the majority of people living in rural areas.

Aizawl is the capital city of Mizoram, India. The location is situated atop a peak of 3715 feet above sea level in the northern section of Mizoram. The terrain is mostly mountainous with narrow deep valleys hillocks and steep hills interspersed. The geology of Aizawl is composed of repeated sandstone, siltstone, mudstone, and shell limestone, and the soil is young and moderate to severely acidic.

Serchhip district is located in northern Mizoram and is bordered by Aizawl district on the northwest, Lunglei district on the south, and Champhai district on the east. Myanmar borders the region on the international level in the southeast. Agriculture is the main source of income for the residents of Serchhip district with jhum being the most common cultivation.

Mamit district is located in the western part of Mizoram and is bordered on the north by Hailakandi district Assam, on the west by North Tripura, Tripura, on the south by Lunglei district, and on the east by Kolasib district and Aizawl district, as well as an international border with Bangladesh. The forest and the Dampa Tiger Reserve are two well-known reserve forests in the area. This region boasts of good soil and an agrarian economy. Jhum agriculture has been the primary source of agricultural products.

Table 1: Collection sites of Solanaceae plants species.

Districts	Sample ID	Location	**GPS Map**	
			Latitude	**Longitude**
AIZAWL	A1	Sihphir	23.81°N	92.73°E
	A2	Lengpui	23.94°N	92.35°E
	A3	Hmuifang	23.55°N	92.70°E
	A4	Sialsuk	23.40°N	92.75°E
	A5	Lungleng	23.65°N	92.67°E
SERCHHIP	S1	Chhingchhip	23.68°N	92.85°E
	S2	Maite	23.69°N	92.96°E
	S3	Chhiahtlang	23.37°N	92.84°E
MAMIT	M1	Reiek	23.69°N	92.60°E
	M2	Dampui	23.94°N	92.35°E
	M3	Zawlnuam	23.99°N	92.36°E

Figure 1: Map of Mizoram showing study and collection sites (Courtesy https:www.mapsofindia.com).

3.2. Ethnobotanical studies

3.2.1. Field Survey

Ethnobotanical information possessed by local people of the study sites was collected during the spring and summer of 2017, 2018, and 2019. The fieldwork was based on data collection on ethnobotanical uses. Formal and informal interviews with the informants, the villagers, and the local practitioners and healers were done. Plant parts used and modes of preparation were also recorded. The specimen of each plant claimed to have ethnobotanical values belonging to Solanaceae were collected and identified and were stored at the Herbarium of the Department of Botany, Mizoram University, Aizawl, Mizoram, India.

In my study, the personal information such as age, gender, profession, and education level for each informant was recorded. The informants (traditional healers, farmers, village leaders) were local inhabitants ranging in age from 25 to 85 years old. All the information was acquired after receiving an oral prior informed consent from the informants, according to the International Society of Ethnobiology (ISE) code of ethics. During the interviews, the informants were requested to specify plants' vernacular name, plant part used, traditional uses, and method of preparation. Mostly, the interviews took place in the informants' houses or their farms (Figure 2).

Figure 2: Field survey, sample collection, and semi-structured interviews with local inhabitants.

3.2.2. Specimen identification

The field observations with informants were conducted to identify the morphological traits and habitats of each therapeutic plant. Voucher specimens and photographs were identified and confirmed according to various botanical websites (Eg., www.theplantlist.org, www.worldfloraonline.org, www.kew.org.) Voucher specimens were made and deposited in the Herbarium of the Department of Botany, Mizoram University for future references.

3.2.3. Quantitative Indices

The Informant Consensus Factor (F_{ic}) which was derived for each medicinal category was used to quantify the species' use variability (Trotter and Logan 1986) and was calculated as:

$$F_{ic} = (N_{ur} - Nt)/ (N_{ur} - 1)$$

Where,

N_{ur} is the number of uses record from the informants for a particular plant use category

Nt is the number of species used for each ailment mentioned by the informants.

The Informants Consensus Factor value ranges from 0 to 1, where the high value of F_{ic} represents an agreement among the informants on the use of species in a medicinal subcategory.

The Fidelity Level index (FL) was used to determine the plant species that was preferred by the informants to treat a particular disease (Friedman et al. 1986) and was calculated as follows:

$$FL\ (\%) = (Np/N)\ X\ 100$$

Where,

NP is the number of informants who reported the use of species for the treatment of a particular disease.

N is the number of informants who used the plants as medicine to treat any given disease.

3.3. Molecular Study

3.3.1. DNA Isolation

The total genomic DNA was extracted from young leaves of collected samples using the modified CTAB method. Firstly, 1g of young leaf was washed with sterile distilled water and grounded using mortar and pestle in 500μl extraction buffer (100mM Tris HCl, 1.5M NaCl, 20mM EDTA, 2% CTAB, 1% PVP at pH 8), and incubated at 60°C for 30 mins followed by centrifugation at 11,000rpm for 15mins. The supernatant was transferred to a new tube and 10μl RNaseA was added to it and then incubated at 30°C for 30 mins. Then, 500μl of Chloroform Isoamyl was added to the samples followed by centrifuge at 11,000rpm for 1min. Then, 500μl of ice-cold isopropanol was added to the tube and kept at -20°C for 1hr. The sample was centrifuged at 11,000rpm for 10 mins. The tube containing DNA was washed with 70% ethanol and the DNA was then air-dried. DNA was dissolved in 30μl TE (10mM Tris HCl, 1mM EDTA).

Procedure:

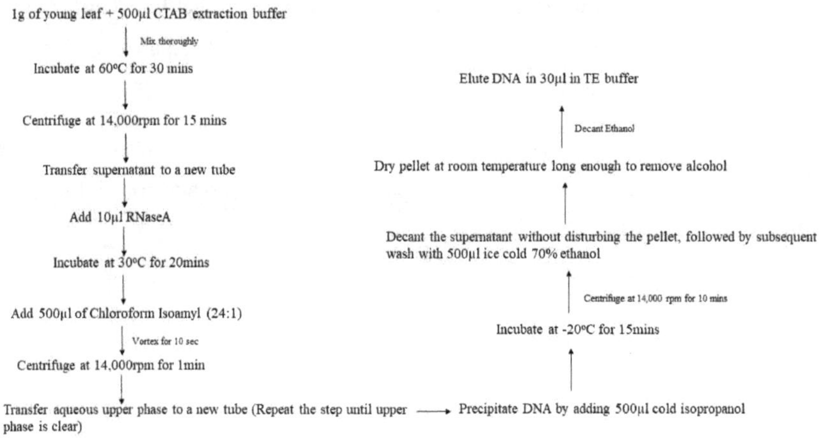

Then, quality of the DNA was checked by standard 0.8% agarose gel electrophoresis and UV-VIS spectrophotometry.

3.3.2. PCR Amplification

PCR amplification was performed in a 25 μl reaction volume containing 1X PCR buffer with MgCl2, 0.25M dNTPs, 0.25 pm each primer, 1U Taq polymerase (Takara Japan) in a Mastercycler Nexus Gradient Thermocycler (Eppendorf AG, Hamburg, Germany) with the parameter settings of 35 cycles of denaturing at 95°C for 1 min, annealing for 1 min and extension at 72°C for 1 min. Two sets of genes – *rbcL* and *ITS2* (Table 2) were used to amplify the DNA. The amplified products were electrophoresed on a 1.5% agarose gel and then quantified spectrophotometrically. Lastly, the products were cleaned before sending for commercial sequencing to Eurofins Genomics India Pvt Ltd., Bengaluru, India.

Table 2: Primer used for PCR Amplification.

Sl.No	Gene	Primer sequence	Reference
1.	*ITS2*	F - GAAGGAGAAGTCGTAACAAGG R - TCCTCCGCTTATTGATATGC	Taberlet et al. 2007
2.	*rbcL*	F- CTGTATGGACCGATGGACTTAC R-CGGTGGATGTGAAGAAGTAGAC	Taberlet et al. 2007

3.3.3. Sequence Analysis

Each sequence was annotated based on chromatogram and BLAST result and searched with a reference database of all *rbcL* and *ITS2* regions using the NCBI BLAST (https://blast.ncbi.nlm.nih.gov/Blast.cgi). The sequence was computed to assign identity and the ID of each species was associated with the best BLAST hit and E-value cut off. This corresponds to choosing the top hit in the BLAST results (BLAST1). The phylogenetic analysis based on *ITS2* and *rbcL* regions was performed using MEGA 7.0 (Tamura et al. 2007) with neighbor-joining (NJ) method and 1000 bootstrap replications.

3.4. Phytochemical Study

3.4.1. Samples preparation

A 50g of the edible parts of powdered samples was extracted with 500mL of methanol using the Soxhlet apparatus for 25 cycles (Figure 3). The extracts were then concentrated using a water bath (50°C) until it formed a paste. Each sample was finally made a concentration of 100μg/mL using methanol.

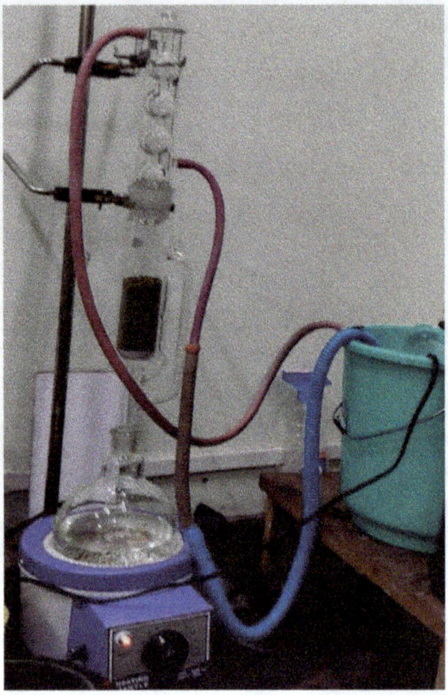

Figure 3: Extraction of plant specimen using Soxhlet apparatus for phytochemical analysis.

3.4.2. Phytochemical screening

Phytochemicals such as phenol, saponins, flavonoids, tannins, and terpenoids from the methanolic extracts were estimated using the procedure proposed by Nwankwo & Ukaegbu-Obi (2014).

3.4.2.1.Test for Tannins

A 100mg of plant extracts was mixed with 2ml of 2% $FeCl_3$ solution and the appearance of blue-green color indicates the presence of tannins.

3.4.2.2.Test for Flavonoids

About 10 ml of plant extracts was mixed with 2ml of 2% NaOH solution. The yellow color formed which turned colorless after the addition of diluted HCl indicated the presence of flavonoids.

3.4.2.3. Test for Saponins

A total of 10 ml of plant extract was mixed with 4ml of distilled water and shaken vigorously. The formation of foam was observed which indicated the presence of saponins.

3.4.2.4. Test for alkaloids

A 10 ml of plant extracts was mixed with 10ml methanol and filtered and 1% HCl (2 ml) was added to the filtrate and kept for 1 minute. Then, 6 drops of Wagner reagent were added and the brownish-red precipitate was observed which indicated the presence of alkaloids.

3.4.2.5. Test for terpenoids

About 5ml of plant extracts was mixed with 2ml of chloroform and 3ml of H_2SO_4. The reddish-brown color was observed which indicated the presence of terpenoids.

3.4.3. Quantitative Analysis

3.4.3.1. Determination of total Phenol

The total phenolic content of the extracted samples was determined using the Folin-Ciocalteu method following Mc Donald et al. (2001). The plant extract was mixed with Ciocalteu reagent (0.1ml, 1N) and incubated at room temperature for 15 minutes. Then, 5ml of Na_2CO_3 was added and incubated at room temperature for 30 minutes and the absorbance was measured at 760nm. Gallic acid was used as standard and total phenolic content was expressed in terms of gallic acid equivalent (mg g^{-1} of the extracted compound).

3.4.3.2. Determination of total flavonoids (TFC)

Total flavonoid content was determined using the Aluminium chloride colorimetric method (Chang et al. 2002). Brief, 1ml plant extract was mixed with 1ml methanol, 0.5ml aluminium chloride (1.2%), and 0.5ml Potassium acetate (100mM) and incubated at room temperature for 30 mins. Then, the absorbance was measured using a spectrophotometer at 415nm. Quercetin was used as standard and total flavonoid content was expressed in terms of quercetin equivalent (mg g^{-1} of the extracted compound).

3.4.3.3.Determination of Total Anthocyanin content (TAC)

The total anthocyanin content was measured using a method proposed by Abdel-Aal & Hucl (1999). The methanol extracts were mixed with acidified methanol (Methanol and 1N HCl, 85:15 v/v, pH1) and the absorbance was taken at 535nm against reagent blank. Cyanidin 3-Glucoside was used as a standard. Then TAC was calculated as:

$$TAC\ (\mu g/g) = (A/\varepsilon) \times (vol/1000) \times MW \times (1/sample\ wt) \times 10^6$$

Where A is absorbance, ε is molar absorptivity of Cyanidin 3-Glucoside, vol is the total volume of anthocyanin extract and MW is the molecular weight of Cyanidin 3-Glucoside.

3.5. Enzymatic Antioxidant Assay

3.5.1. Determination of Catalase (CAT) activity

CAT activity was determined following Sunohara and Matsumoto (2004). Briefly, 0.1ml of the extract was mixed with 1.9ml of 25 mM H_2O_2 in 50mM potassium phosphate buffer (pH 7). Then, the absorbance was measured at 240nm. The enzyme activity was defined as the amount of H_20_2 (mM) decomposed per minute.

3.5.2. Determination of Ascorbate peroxidase (APX) activity

APX activity was determined by using Sunohara and Matsumoto's (2004) method with some modifications. About 2ml of extract was mixed with 0.5ml of 100mM potassium phosphate buffer (pH 7), 0.5ml of 1mM ascorbic acid, 0.5ml of 0.4mM EDTA and 0.02ml 0f 10mM H_20_2. Then, absorbance was measured at 290nm. The enzyme activity was defined as the number of H_2O_2 (mM) decomposed per minute.

3.5.3. Superoxide dismutase (SOD) activity

The SOD activity was determined following McCord (2001). About, 3 ml of the extract was mixed with 1.5M sodium carbonate, 0.1 ml of 3 mM EDTA, 0.2 ml of 200 mM methionine, 0.1 ml of 2.25 mM NBT, 1.5 ml of 100 mM potassium phosphate buffer, 0.95 ml of distilled water, and 0.5 ml of extract. The tube without the extract was taken as a control. The reaction was started by adding 0.1 ml riboflavin (60uM) under light for 15 mins. Then, the absorbance was measured at 560 nm and 1 unit of enzyme activity was defined as the quantity of enzyme which reduced the absorbance reading of samples by 50% in comparison with the control.

3.5.4. 2,2-diphenyl-1-picrylhydrazyl (DPPH) radical scavenging activity

Antioxidant activity of the extract was determined with DPPH radical scavenging method (Yan-Hwa et al. 2000). To 50μl of 10-100μg/mL plant extract, 2 ml DPPH was added and kept in dark at room temperature for 30 mins. Then, 1 ml methanol and 2 ml DPPH were used as positive control while methanol solution was used as a negative control. Then, the absorbance was measured at 517 nm. The percentage of DPPH radical scavenging activity (%RSA) was calculated as:

%RSA = 100 X (absorbance of control – Absorbance of the sample)/ Absorbance of control

3.6. Nutritional Study

3.6.1. Proximate analysis

For the estimation of protein and carbohydrates, 500 mg of edible parts were homogenized with phosphate buffer (50mM, pH 7.6). The extract was centrifuged at 8000 rpm for 10 min at 4°C. The supernatant was then used for estimation of protein content following Lowry's method (Lowry et a. 1951) and Carbohydrate content using Hall's (2007) method (Anthrone reagent) with glucose as a standard.

3.6.2. Determination of mineral ion content

One gram of air-dried sample was crushed and digested using Nitric acid (HNO_3) and Hydrogen Peroxide (H_2O_2) in a 5:1 ratio until it became crystal clear. The clear sample was cooled and diluted with distilled water to make up to 50 ml. The diluted solution was then filtered using a 0.2-micron membrane filter and analyzed for detection of elements using Atomic Absorption Spectroscopy (Shimadzu AA-7000, Japan) and Microwave Plasma Atomic Emission Spectroscopy (4100 MP-AES, Agilent Technologies, USA).

3.7. Antibacterial Study

3.7.1. Bacterial strains

In vitro antimicrobial activity was tested for methanolic extracts from 10 Solanaceae species against three bacterial strains viz. *Bacillus subtilis* ATCC11774, *Pseudomonas aeruginosa* ATCC9027 and *Escherichia coli* ATCC1229.

3.7.2. Media preparation and Antibacterial activity

The antimicrobial assay was performed using the agar well diffusion method in a nutrient broth agar plate. A 40μl each of the extract was added to the well. The plate was incubated overnight at 30^0C and then the diameter of the zone of inhibition was measured. All the experiments were performed in triplicates using appropriate positive controls.

3.8. FT-IR Analysis

Functional groups present in the studied samples were analyzed using Fourier transformed infrared (FT-IR) spectroscopy for the frequency ranging from 400-4000 cm^{-1}.

3.9. Statistical analysis

All the results were reported as the mean ± standard deviation. The linear regression coefficient (R^2) for total flavonoid and phenolic content with antioxidant activity was analyzed using Graph Pad Prism Version 5. P-value < 0.05 was considered significant.

Chapter 4

Results

4.1. Collection of Solanaceae plants

Solanaceae plants collected were recorded and deposited to the Herbarium of the Department of Botany, Mizoram University, Aizawl, Mizoram, and the voucher numbers of the samples were collected. In my study, 20 species (Photo plates 1 to 5) belonging to Solanaceae were collected from the study sites.

Brugmansia suaveolens

Capsicum annuum

Capsicum frutescens

Cestrum nocturnum

Photo plate 1

Datura metel

Lycopersicon esculentum

Lycianthes neesiana

Nicotiana tabacum

Photo plate 2

Petunia axillaris

Physalis angulata

Physalis peruviana

Solanum americanum

Photo plate 3

Solanum anguivi

Solanum betaceum

Solanum incanum

Solanum melongena

Photo plate 4

Solanum pseudocapsicum

Solanum tuberosum

Solanum torvum

Solanum viarum

Photo plate 5

4.1.1. Description of Solanaceae plants

1. *Brugmansia suaveolens* (Humb & Bonpl. Ex. Willd)Bercht & J. Presl
Common Name: Angel's Trumpet

Local name: Tawtawrawt par

Habit: Woody Shrub

Flowering time: February – July

Flower Color: White

Woody shrub, single trunk with many-branched, alternate leaves with entire toothed margin, oval and pointed. Shade-loving plant leaves become larger when grown in shade. Flowers trumpet-shaped, pendulous, facing downside. Corolla white, tube greenish towards the base. White, cream, yellow, orange, or pink flower colors are obtained using hybridization. Fruit berry-like capsule. Seeds many with irregular shaped. Found in disturbed areas, riverbanks, and urban open land, abundant in cool and humid areas.

2. *Cestrum nocturnum* Lam.
Common Name: Night-blooming jasmine

Local name: Zan-par rimtui

Habit: Woody Shrub

Flowering time: December – April

Flower Color: White

Flexuous branches, leaves lanceolate-elliptic, the apex is acuminate, base rounded. Flowers spicate with congested racemes, terminal leafy panicles, calyx campanulate; corolla vespertine, producing strongly sweet scent at night, greenish-yellow color with tubular, slender and enlarged towards the apex. Berries white with hard or juicy form; Seeds are less. Abundant in moist and open disturbed areas.

3. *Capsicum annuum* L.
Common Name: Bell pepper

Local name: Hmar-cha

Habit: Shrub

Flowering time: July - September

Flower color: White

Multiple branched annual plants. Leaves are simple, alternate, entire leaves margin, light to dark green color. Flowers are normally borne individually, terminal. The calyx is cup-shaped with 5 conspicuous toothed. Corolla is campanulate with white color. Fruits are non-pulpy berries, different in size, shape, color, and degree of pungency, conical-shaped with green, yellowish, creamy, or purplish when the immatured stage, in mature stage the color change to red, orange, yellow, brown. Seeds circular flattened with pale yellow. It can grow in dry, sub-arid, and humid places.

4. *Capsicum frutescens* L.

Common Name: Chilli pepper

Local name: Hmar-cha vankawk

Habit: Shrub

Flowering time: July - September

Flower color: Pale Yellow

Multiple branched annual plants. Flowers pale yellow. Fruits grow erect with conical to lanceoloid shaped and very small and pungent, pale yellow in immature and changes to bright red in a mature stage. Grow in dry, sub-arid, and humid places.

5. *Datura metal* L.

Common Name: Jimson weed/Devil's trumpet

Local name: Datura

Habit: Shrub

Flowering time: February-April

Flower Color: Purplish with white

Branched with few hairs. Leaves are ovate to lanceolate, entire margin. Flowers independent in the stem, calyx angular to tubular, corolla white, purplish, or yellowish. Capsule globose to ovate, spiny. Seeds numerous, yellow-brown, flattened. Grown as ornamentals with fertile soil.

6. *Lycianthes neesiana* Wall. Ex Nees.

Common Name: Nees nightshade

Local name: Vani-an

Habit: Shrub

Flowering time: June – August

Flower color: White

Perennial shrub with velvety-hairy plants. Single trunk with multiple branches. Leaves unequal paired, lanceolate or elliptic, sharp minor leaves. Flowers single, white color with star-shaped. Calyx cup-shaped. Fruits are rounded, small, red to orange in color. Found in dry and hilly places.

7. *Nicotiana tabacum* L.

Common Name: tobacco

Local name: Vaihlo

Habit: Perennial herb

Flowering time: June – October

Flower Color: Pinkish-white

Annual branched perennial herbs and all plant parts are sticky, covered with short glandular hair. Leaves are simple, alternate, large with short-stalked, green color, elliptic or ovate shaped. Flowers inflorescence, triangular lobes, pinkish-white with light violet color, tips yellowish white. Seeds are small, many, ovoid or bean-shaped, brownish. Found in moist places, waste sides, shady places.

8. *Petunia axillaris* Lam.

Common Name: Petunia

Local name: Petunia

Habit: Herb

Flowering time: December – March

Flower color: White

Annual herbaceous plants. Leaves are hairy, sessile, oval-shaped with a smooth margin. Flowers trumpet-shaped and obtained different colors by hybridization. Seeds are minute covered in a capsule. The sun-loving plant does not tolerate shade.

9. *Physalis angulata* L.

Common Name: Cutleaf groundcherry, wild gooseberry

Local name: Chalpangpuak/Kelasai-raw-phit

Habit: Herb

Flowering time: August-October

Flower color: Yellowish-Green

Annual, erect branched herbaceous, glabrous plants with tap-roots. Leaves ovate or lanceolate dark green color and bases are unequal. Flowers yellowish-green without spots. Berries are rounded with a protective cover, light green in unripe stage, and orange color in ripe. Calyx pale green and. Seeds are flat, disc-shaped with pale yellow.

10. *Physalis peruviana* L.

Common Name: Cape gooseberry

Local name: Pangpuak chi khat

Habit: Herb

Flowering time: August – October

Flower color: Yellowish with a dark spot in the middle.

Annual, branching herbaceous, all parts of plants are covered with glandular hair. Leaves are silky, randomly toothed with heart-shaped, greenish color, after-ripening leaves turn yellowish and fall off. Flowers are hermaphrodite, bell-shaped, yellow with a dark purple-brown spot in the throat. The calyx is purple-green, veined. Berries are round, smooth orange-yellow, glossy skin with lots of juice. Seeds are numerous and yellowish. Grown in sandy soil with much sunlight.

11. *Solanum americanum* Mill.

Common Name: American Black nightshade

Local name: Anhling

Habit: Herb

Flowering time: August-February

Annual or perennial herb, single trunk, erect with multiple branches. Stem rounded, glabrous. Leaves are sharp ovate to lanceolate. Flowers regular, lobes ovate to oblong, stellate, white or flushed purple with basal yellow-green. Anthers yellow, stigma pale green. Fruits are shiny globose, green glossy turning to purplish-black at maturity. Seeds are numerous, discoid, creamy in color. Grown in roadside and disturbed places.

12. *Solanum anguivi* Lam.

Common Name: Forest bitter berry/ African eggplant

Local name: Take

Habit: Woody Shrub

Flowering time: Throughout the year

Flower color: Purple

Semi-woody shrub and all parts of plants are covered with hair and thorn. Stem thick purplish-yellow. Leaves are ovate, prickly hairy, and light-greenish color. Flowers stellate, hairy purple color. Fruits are glabrous, green color in immature and yellow to light red after ripening with shiny globose. Grown as home-garden, roadside, and disturbing places with much sunlight.

13. *Solanum betaceum* Cav.

Common Name: Tree tomato/ Tomarillo

Local name: Thing tomato

Habit: Woody Shrub

Flowering time: June-September

Flower color: Pinkish white

Woody shrub, single trunk with many-branched plants. Leaves are large, simple, and perennial with a strong pungent smell. Flowers form clusters and are pink-white. Fruits are oval-shaped, light greenish during immature and yellowish and orange to red color in a mature stage with dark horizontal stripes. Grown in light, deep and fertile soil.

14. *Solanum incanum* L.

Common Name: Bitter tomato

Local name: Samtawk

Habit: Shrub

Flowering time: August – November

Flower Color: White

Perennial shrub covered with velvety glandular hair and spines all parts of plants. Leaves alternate with oval-shaped with waxy margin and light-greenish color. Flowers are borne in leaf axils, white sometimes pale purple. Fruits are ovoid or fusiform, globose, red to orange in color. Seeds numerous, flattened, pale-yellow, or brown color. Grown as cultivated plants in lands and home gardens.

15. *Solanum lycopersicon* L.

Common Name: Tomato

Local name: Tomato

Habit: Herb

Flowering time: October – January

Flower Color: Yellow

Annual herb plants with vines and decumbent, branched stem, hairy. Leaves are regular, hairy, serrated margin. Flowers are yellowish with 5 pointed lobes on the corolla. Berry red or orange-yellow with juicy and shiny. Seeds are straw color. Grown in moist, disturbed areas. Abundant in damp roadside places.

16. *Solanum melongena* L.

Common Name: Eggplant

Local name: Bawkbawn

Habit: Shrub

Flowering time: August – November

Flower color: Purple

Perrenial shrub covered with velvety glandular hair. The stems are spiny. Leaves are stellate, light to dark greenish. Flowers are white to purple with 5 lobed corollas, stamens yellow. Fruits are globose to oblong, shiny, and smooth with multiple seeds. Seeds are kidney-shaped with pale brown color. Cultivated as vegetables in lands and home-garden.

17. *Solanum pseudocapsicum* L.

Common Name: Jerusalem Cherry/Winter cherry

Local name: Hmarcha suak

Habit: Shrub

Flowering time: April – October

Flower color: White

Erect, shrub, glabrous plants. Young branches are greenish while older branches turn woody and brownish. Leaves are borne on stalks, oval to lanceolate with a slightly wavy margin. Flowers are angular, white in color, fused at the base. The stamen is yellow. Fruits are rounded, greenish turn to orange-red at maturity, with multiple seeds. Seeds are white or pale yellow. Grown as ornamentals.

18. *Solanum torvum* Sw.

Common Name: Turkey berry

Local name: Tawkpui

Habit: Woody Shrub

Flowering time: November – April

Flower color: Creamy-White

Woody and spiny shrub. Leaves are ovate with blunt lobes, covered with stellate hair. Flowers occur laterally, hairy sepals, 5 white to cream color petals, stamens yellow color. Fruits are globular, green when immature, and turn to pale yellow when mature. Seeds are numerous, flattened, and in red color.

19. *Solanum tuberosum* L.

Common Name: Potato

Local name: Alu

Habit: Herb

Flowering time:

Flower color: White

Erect annual herb, angular, branched stem. Leaves alternate, petiolate and ovate. Flowers white with greenish-yellow central star. Calyx campanulate. Fruits or tubers are sub-globose, yellow-green. Seeds are numerous, flattened, subcircular to ovate, pale yellow-brownish color, and poisonous. Grown as vegetables.

20. *Solanum viarum* Dunal.

Common Name: Tropical soda apple/ Sodom apple

Local name: At-hlo hling

Habit: Shrub

Flowering time: November – February

Flower color: White

Erect and annual shrub with short stems and multiple branches and spiny plants. Leaves are ovate, blunt lobed, dark green, and glossy surface, downside is pale green. Corolla white, anthers pale yellow. Fruits are globose, green spotted with creamy white when immature, and turn into light yellow when mature. Seeds are many, flattened, brownish color. Grown in disturbed places and roadsides.

4.2. Ethnobotanical Studies

The present study showed useful information about the ethnobotanical uses of Solanaceae plants in the study area. The traditional knowledge and ethnobotanical uses of plants among the Mizo people require not only preservation but also documentation for future generations. This study can provide useful information on Solanaceae plants of Mizoram that are used for different purposes. It was observed that 20 species from Solanaceae were used to treat various ailments and the species are detailed in Table 3. Most of the species were collected from the wild while few were found to be cultivated. In total, there were 62 informants and among them 38 were males and 24 were females. The age distribution of the informants ranged from 25-85 years (Table 4). The distribution of the number of informants was highest in the age group 56-65 in males and 36-45 in females (Figure 4).

Table 3: List of collected Solanaceae.

Sl No	Species Name	Local Name	Habit	Part Used	Uses
1	*Brugmansia suaveolens* (Willd) Sweet. (MZU/BOT/101)	Tawtawrawt par	Woody Shrub	Leaves, flower	Dried leaves and flowers are smoked for asthma, used as drugs, induce hallucination, narcotic, psychoactive.
2	*Capsicum annuum* L. (MZU/BOT/102)	Hmarcha te	Shrub	Leaves, Fruits	Fruits used as condiments, spices, improves digestion. Leaves are prepared with fermented pork eaten as vegetables. Fruits and leaves juices are applied to burn and snake bites. Fruits are used as anti-haemorrhoidal, antiseptic, anti-rheumatic.
3	*Capsicum frutescens* L. (MZU/BOT/103)	Hmarchapui	Shrub	Leaves, Fruits	Fruits used as condiments, spices, improves digestion. Leaves are prepared with fermented pork eaten as vegetables. Fruits leave juices applied to burn and snake bite.
4	*Cestrum nocturnum* L. (MZU/BOT/104)	Zan par rimtui	Woody shrub	Leaves, flower	Extracted plants are used as antispasmodic. Used for insect repellent. Also used as ornamentals.
5	*Datura metel* L. (MZU/BOT/105)	Datura	Shrub	Whole plant	Leaves are used for skin diseases, rash, sores, inflammatory swelling, asthma, bronchitis. Leaves and seeds are used as hallucinogenic. Whole plants are used as pesticides and snake repellence.
6	*Lycianthes neesiana* Wall. Ex Nees	Vani an	Shrub	Leaves	Leaves are eaten as vegetables, boiled with

	(MZU/BOT/106)				water. Boil leaves are used as stomached, stress relieve.
7	*Lycopersicon esculentum* Mill. (MZU/BOT/107)	Tomato		Fruits, Leaves	Fruits are eaten as raw or cooked, also used as juice. Fruits are used as skincare, treatment for sunburn. Leaves ground in powder form is applied on spotted skin or leprosy spots.
8	*Nicotiana tabacum* L. (MZU/BOT/108)	Vaihlo	Herb	Leaves	Leaves are mainly used for smoking, sedative, diuretic, antispasmodic. Leaves juices used for insect sting and skin disease.
9	*Petunia axillaris* Lam. (MZU/BOT/109)	Petunia	Herb	Whole plant	Ornamental.
10	*Physalis angulata* L. (MZU/BOT/110)	Chal pangpuak	Herb	Fruits, Leaves	Fruits are eaten raw or cooked. Leaves are used as analgesic, antiseptic, asthma, diarrhea. Fruits are used for the treatment of malaria, liver ailment, rheumatism, indigestion.
11	*Physalis peruviana* L. (MZU/BOT/111)	Chal pangpuak chikhat	Herb	Fruits	Fruits are eaten raw. Fruits are used as antidiuretic, rheumatism, anthelmintic, stomach pain.
12	*Solanum americanum* Mill. (MZU/BOT/112)	Anhling	Herb	Whole plant	Young shoot and leaves eaten as cooked. Decoction of whole plants is used as antispasmodic, anti-inflammatory blood purification, ulcers, anti-cancer, skin disease.
13	*Solanum anguivi* Lam. (MZU/BOT/113)	Tawkte	Woody Shrub	Fruits, root, Leaves	Green fruit is eaten as cooked or raw. Leaves are ground and applied on skin

					disease, rash, and spots. Fruits are used as medicine for high blood pressure, asthma and stomach ache. Roots grounded to powder used as toothache, insect bites.
14	*Solanum betaceum* Cav. (MZU/BOT/114)	Thingtomato	Woody Shrub	Fruits, Leaves	Fruits are eaten as raw, cooked/roasted as vegetables. Also used in inflammatory painful disease, tonsils problem, liver problem. Leaves are heated on low flame and wrapped around the neck for sore throat.
15	*Solanum incanum* L. (MZU/BOT/115)	Samtawk	Shrub	Fruits, roots, snake bites, and wounds.	Green fruits are eaten as cooked or raw. Fruits are used as an analgesic, medicine against high blood pressure, menstrual problem, sore throat, stomach ache, liver problem, rheumatism, conjunctivitis. Roots or fruit rubbed on gums for toothache.
16	*Solanum melongena* L. (MZU/BOT/116)	Bawkbawn	Shrub	Fruits, Leaves, roots	Fruits cooked or roasted. Fruits are used for lowering blood cholesterol levels, high blood pressure, an anti haemorrhoidal, antidote to poisonous mushrooms. Leaves as narcotics, skin disease, treatment for burns and bites. Decoction of leaves and roots used as

					toothache, bleeding and antiasthmatic.
17	*Solanum pseudocapsicum* L. (MZU/BOT/117)	Hmarcha suak	Shrub	Whole plant	Mainly used for ornamental. Used for insect repellent.
18	*Solanum torvum* Sw. (MZU/BOT/118)	Tawkpui	Woody Shrub	Fruits	Young fruits cooked or raw. Fruits are used for the treatment of fever, sore throats, stomach aches, chest pain. Used as antidiuretic, antidiabetic.
19	*Solanum tuberosum* L. (MZU/BOT/119)	Alu	Herb	Leaves, Tubers	Tubers are eaten as vegetables. Leaves are used for antispasmodic in chronic cough. Juices/grounded tubers are used for the treatment of peptic ulcers and also applied to burns.
20	*Solanum viarum* Dunal. (MZU/BOT/120)	At-hlo hling	Shrub	Fruits, Seeds	Fruits are used for the treatment of rheumatic, arthritis, chronic asthma. Ashes of fruits and seeds are used as a toothache.

Table 4: Demographic characteristics of local informants.

Distribution	No. of Informants
Informants	62
Male	38
Female	24
Total	62
Age range	25-85

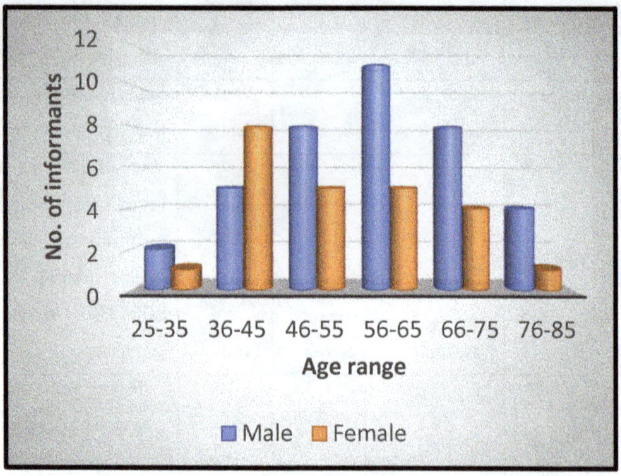

Figure 4: Demographic data of the local informants.

4.2.1. Habit and plant parts used

The growth form analysis of Solanaceae plants revealed that shrubs constitute the highest proportion, representing by 45%, followed by herb 30% and woody shrub 25% (Fig. 5).

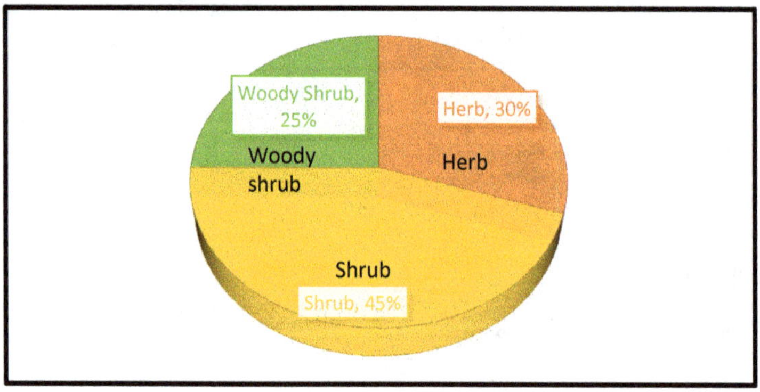

Figure 5: The growth form (habit) of medicinal plants documented and collected.

It was observed that local people use different plant parts for the preparation of traditional medicines eg., leaves, barks, seeds, flowers, fruit, tuber. Leaf was found to be the most used plant part followed by fruit, whole plant, root, flower, seed, and tuber (Figure 6).

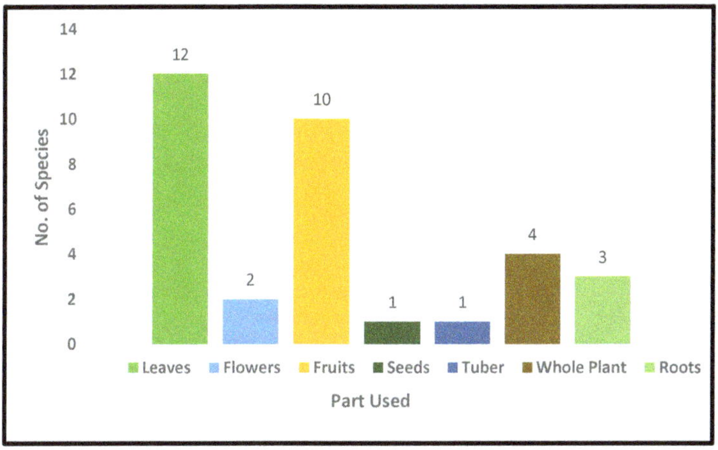

Figure 6: Plant parts used for the traditional treatment of various ailments.

4.2.2. The Informant Consensus Factor

The informant consensus factors have been calculated for each disease category. The highest value (1) of informant consensus factor was obtained for the diseases related to a snake bite, sedative, analgesic, diarrhea, malaria, cancer, and diabetes and the least one (0.84) was associated with asthma (Figure 7).

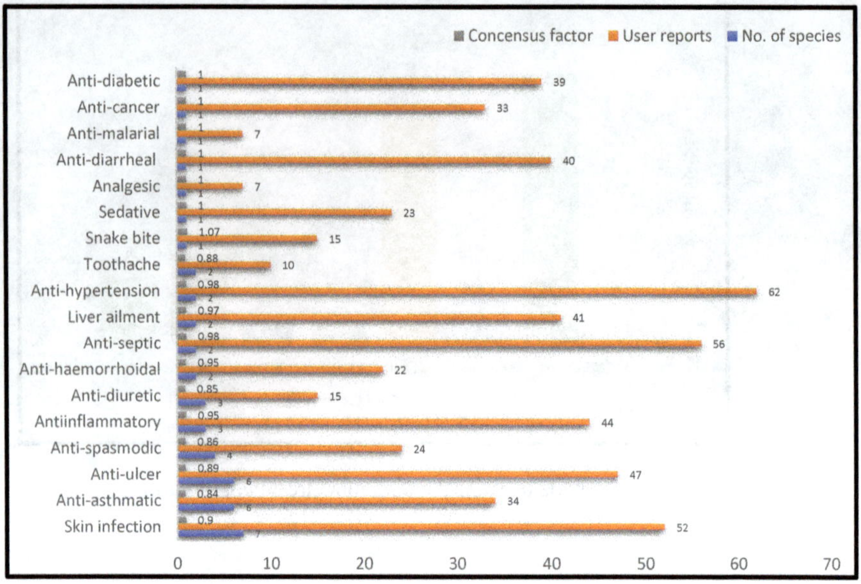

Figure 7: Informant consensus factor (F_{ic}) of Solanaceae plants used for the treatment of various diseases.

4.2.3. The fidelity level index

The fidelity level indices of Solanaceae plants reported against various diseases were calculated (Table 5) and the highest fidelity value (100%) was obtained from the species *S. anguivi* used for the treatment of hypertension and the least one (30%) from *S. tuberosum* which are used for the treatment of ulcer.

Table 5 Fidelity level values of medicinal plants reported against various diseases categories.

Ethnomedicinal Plants	**Diseases**	**NP**	**N**	**FL (%)**
Brugmansia suaveolens	Anti-asthmatic	12	17	70.58
Datura metel		18	21	85.71
Physalis angulata		13	16	81.25
Solanum anguivi		10	14	71.42
Solanum melongena		14	19	73.68
Solanum viarum		11	15	73.33

Capsicum annuum	Anti-hemorrhoidal	21	24	87.5
Solanum melogena		16	24	66.6
Capsicum annuum	Anti-septic	18	20	90
Physalis angulata		9	12	75
Capsicum annuum	Anti-rheumatic	21	22	95.45
Physalis peruviana		9	11	81.8
Solanum viarum		12	25	48
Solanum incanum		18	31	58.06
Solanum tuberosum	Skin infection	13	17	76.47
Capsicum frutescens		10	17	58.82
Datura metel		16	21	76.19
Lycopersicon esculentum		13	17	76.47
Nicotiana tabacum		29	31	93.54
Solanum americanum		11	18	61.1
Solanum melongena		21	28	75
Capsicum frutescens	Snakebite	8	12	66.6
Cestrum nocturnum	Anti-spasmodic	8	10	80
Nicotiana tabacum		18	21	85.71
Solanum americanum		16	19	84.21
Solanum tuberosum		13	18	72.22
Datura metel	Anti-inflammatory	11	13	84.61
Solanum americanum		20	21	95.23
Solanum betaceum		4	11	36.3
Datural metel	Bronchitis	5	9	55.55
Lycianthes neesiana	Anti-ulcer	11	15	73.33
Physalis angulata		4	10	40
Physalis peruviana		4	6	66.6
Solanum americanum		10	20	50
Solanum anguivi		15	26	57.69
Solanum incanum		12	15	80

Solanum torvum		9	11	81.8
Solanum tuberosum		3	10	30
Nicotiana tabacum	Anti-diuretic	16	22	72.7
Physalis peruviana		4	9	44.44
Solanum torvum		11	19	57.89
Nicotiana tabacum	Sedative	21	30	70
Physalis angulata	Analgesic	5	9	55.55
Physalis angulata	Antidiarrheal	5	15	33.33
Physalis angulata	Anti-malarial	4	14	28.57
Physalis angulata	Liver ailment	9	21	75
Solanum betaceum		17	25	68
Solanum americanum	Anticancer	10	21	47.61
Solanum anguivi	Antihypertension	32	32	100
Solanum incanum		29	34	85.29
Solanum torvum	Anti-diabetic	6	11	54.54
Solanum viarum	Toothache	18	29	62.06
Solanum incanum		10	16	62.5

Where NP = Number of informants that reported the use of plants for the treatment of a particular disease; N= Number of informants who used the plants as a medicine to treat any given disease; FL= Fidelity Level.

4.3. Molecular Studies

4.3.1. DNA Isolation, PCR amplification, and sequence analysis

The genomic DNA of the collected species was successfully isolated (Figure 8). Amplification of the gDNA with four gene loci- *rbcL, ITS2, matK,* and *rpoC1* was also carried out. Among them, *rbcL* and *ITS2* showed 100% PCR amplification success rates (Table 6) and hence they had been used in my study (Figure 9). All the sequences have been submitted to the NCBI Genbank (Table 7 and 8). It was observed that the *rbcL* region of *Solanum torvum* and the *ITS2* region of *Solanum nigrum, Solanum pseudocapsicum, Solanum torvum, Solanum viarum* showed 100% sequence similarity with the reference sequences (Table 7 and 8). The sequence recovery success rates were also very high for *rbcL* (100% species, 100% genus) and *ITS2* (100% species, 100% genus) among the primers used. And the *matK* and *rpoC1* regions showed very low recovery rates (30% species and 54% genera) (Table 6).

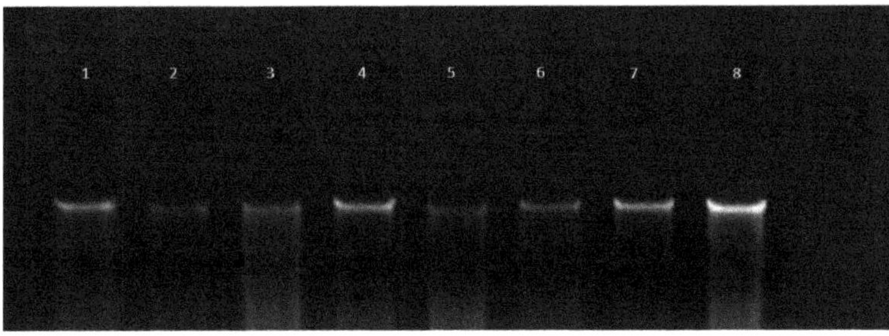

Figure 8: A 0.8% agarose gel of the isolated DNA (1. *B. suaveolens*; 2. *C. annuum*; 3. *C. frutescens*; 4. *D. metel*; 5. *P. angulata*; 6. *P. peruviana*; 7. *L. esculentum*; 8. *S. anguivi*).

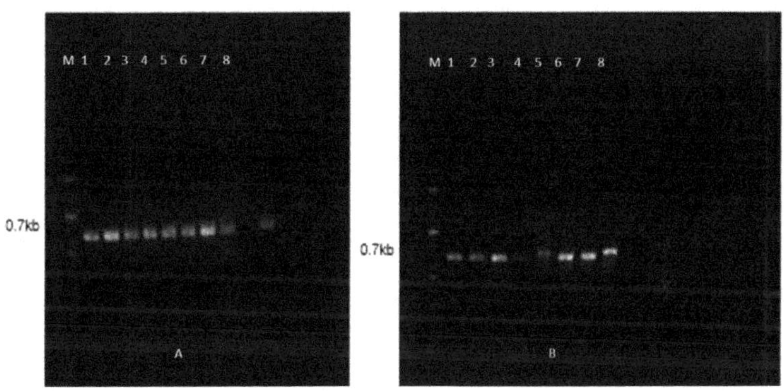

Figure 9: 1.5% agarose gels showing amplified products of *rbcL* (A) and *ITS2* (B). (M – 100bp DNA marker; 1. *B. suaveolens*; 2. *C. annuum*; 3. *C. frutescens*; 4. *D. metel*; 5. *P. angulata*; 6. *P. peruviana*; 7. *L. esculentum*; 8. *S. anguivi*).

Table 6: Sequence recovery rates for four DNA barcodes loci.

Barcode region	*rbcL*	*ITS2*	*matK*	*rpoC1*
Successful species/sample species	20/20(100%)	20/20 (100%)	5/20 (25%)	9/20 (45%)
Successful genera/sample genera	20/20(100%)	20/20 (100%)	8/20 (40%)	12/20 (60%)

Table 7: Solanaceae used with rbcL region, identification using BLASTn and Accession Number

Sl. No	**Species**	**Genome region**	**Taxonomic level**	**BLAST similarity**	**Sequence cover**	**E-value**	**Accession No.**
1.	*Brugmansiasuaveolens*	rbcL	Species	99.69%	98%	0.0	MK984174
2.	*Capsicum anuum*	rbcL	Species	99.84%	97%	0.0	MK984176
3.	*Capsicum frutescens*	rbcL	Species	99.72%	95%	0.0	MK984172
4.	*Cestrum nocturnum*	rbcL	Species	99.37%	96%	0.0	MK984163
5.	*Datura metel*	rbcL	Species	99.69%	97%	0.0	MK984162
6.	*Lycopersicon esculentum*	rbcL	Species	99.45%	98%	0.0	MK984171
7.	*Nicotiana tabacum*	rbcL	Species	99.84%	98%	0.0	MK984177
8.	*Lycianthes neesiana*	rbcL	Species	94%	99.38%	0.0	MK958809
9.	*Petunia axillaris*	rbcL	Species	84%	98.95%	0.0	MK573903
10	*Physalis angulata*	rbcL	Species	99.69%	96%	0.0	MK984165
11	*Physalis peruviana*	rbcL	Species	98.62	94%	0.0	MK984168
12	*Solanum anguivi*	rbcL	Species	99.53%	95%	0.0	MK984173
13.	*Solanum melongena*	rbcL	Species	99.53%	98%	0.0	MK984166
14.	*Solanum americanum*	rbcL	Species	98.37%	99%	0.0	MK984170
15.	*Solanum betaceum*	rbcL	Species	99.41%	98%	0.0	MK573909
16.	*Solanum pseudocapsicum*	rbcL	Species	99.89%	93%	0.0	MK984164
17.	*Solanum incanum*	rbcL	Species	99%	99.4%	0.0	MK573920
18.	*Solanum torvum*	rbcL	Species	100%	98%	0.0	MK984161
19.	*Solanum tubersom*	rbcL	Species	99.84%	96%	0.0	MK984167
20.	*Solanum viarum*	rbcL	Species	99.45%	99%	0.0	MK984162

Table 8: Solanaceae used with ITS2 region, identification using BLASTn and Accession Number

Sl. No	Species	Genome region	Taxon-omic level	BLAST similarity	Sequence cover	E-value	Accession No.
1.	*Brugmansia suaveolens*	ITS2	Species	99.85%	93%	0.0	MH573914
2.	*Capsicum anuum*	ITS2	Species	96.41%	94%	0.0	MH573916
3.	*Capsicum frutescens*	ITS2	Species	98.15%	96%	0.0	MH573912
4.	*Cestrum nocturnum*	ITS2	Species	99.71%	95%	0.0	MK958809
5.	*Datura metel*	ITS2	Species	99.86%	96%	0.0	MK958812
6.	*Lycopersicon esculentum*	ITS2	Species	99.89%	98%	0.0	MH573911
7.	*Nicotiana tabacum*	ITS2	Species	99.28%	98%	0.0	MH573917
8.	*Lycianthes neesiana*	ITS2	Species	94%	99.08	0.0	MH573919
9.	*Petunia axillaris*	ITS2	Species	90%	99.85%	0.0	MH573902
10.	*Physalis angulata*	ITS2	Species	99.81%	97%	0.0	MH573922
11.	*Physalis peruviana*	ITS2	Species	98.62	94%	0.0	MK958811
12.	*Solanum anguivi*	ITS2	Species	97%	97%	0.0	MH573913
13.	*Solanum melongena*	ITS2	Species	99.40%	99%	0.0	MH573920
14.	*Solanum americanum*	ITS2	Species	100%	100%	0.0	MH573910
15.	*Solanum betaceum*	ITS2	Species	95%	99%	0.0	MH573911
16.	*Solanum incanum*	ITS2	Species	100%	97%	0.0	MH573920
17.	*Solanum pseudocapsicum*	ITS2	Species	100%	94%	0.0	MK958810
18.	*Solanum torvum*	ITS2	Species	100%	100%	0.0	MH549149
19.	*Solanum tubersom*	ITS2	Species	98.6%	94%	0.0	MH573905
20.	*Solanum viarum*	ITS2	Species	100%	100%	0.0	MH573909

4.3.2. Phylogenetic analysis

The phylogenetic relationship of Solanaceae species was analyzed using two DNA barcode loci- *rbcL* and *ITS2*. The Neighbor-Joining (NJ) trees based on *ITS2* and *rbcL* regions showed that the species could be grouped into two main clusters (Figures 10 and 11). Based on *rbcL* sequences, *Solanum anguivi, Solanum melongena, Solanum viarum, Solanum americanum, Brugmansia suaveolens, Datura metel, Physalis angulata, Solanum pseudocapsicum, Solanum torvum, Capsicum annuum, Capsicum frutescens, Lycopersicon esculentum, Solanum tuberosum* were grouped into one cluster and *Cestrum nocturnum* and *Nicotiana tabacum* were grouped under the second cluster. It was also observed that all the members of the genus *Solanum* were closely related. The phylogenetic analysis of the *ITS2* region showed that the genus *Solanum* were under the same clade which showed that they were closely related. Likewise, the genus *Capsicum*, *Physalis* were grouped. Interestingly the ornamental plants *Cestrum* and *Petunia* were under the same clade.

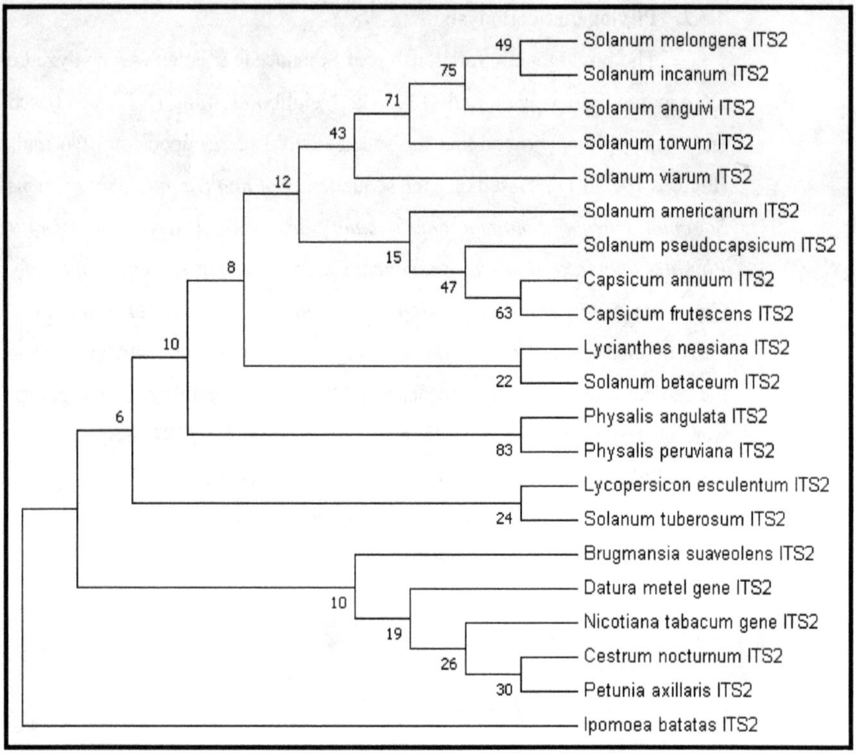

Figure 10: Phylogenetic analysis using a Neighbor-Joining tree of Solanaceae plants of Mizoram using *ITS2* locus.

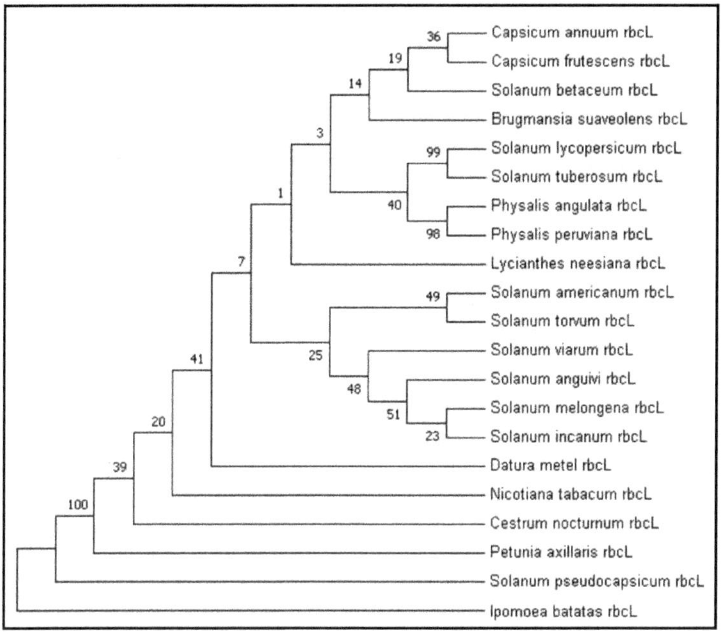

Figure 11: Phylogenetic analysis using a Neighbor-Joining tree of Solanaceae plants of Mizoram using *rbcL* gene loci.

4.4. Phytochemical Studies

4.4.1. Phytochemical screening

Phytochemical analysis of 10 selected species showed that methanolic extracts of Solanaceae species contained tannins, flavonoid, saponin, alkaloids, and terpenoid (Table 9).

Table 9: Phytochemical screening of Selected Solanaceae plants.

Sl No.	Species Name	Parts tested	Alkaloids (Wagner Test)	Flavonoids (Lead AcetateTest)	Saponin (Foam Test)	Tannins (Gelatin Test)	Terpenoids (Salkowski Test)
1	*Capsicum annuum L.*	Fruits	+	+	+	+	+
2	*Capsicum frutescens L.*	Fruits	+	+	+	+	+
3	*Solanum betaceum*	Fruits	+	+	+	+	+
4	*Lycopersicon esculentum Mill.*	Fruits	+	+	+	+	+
5	*Physalis angulata L.*	Fruits	+	+	+	+	+
6	*Solanum americanum Mill.*	Leaves	+	+	+	+	+
7	*Solanum anguivi Lam.*	Fruits	+	+	+	+	+
8	*Solanum incanum L.*	Fruits	+	+	+	+	+
9	*Solanum melongena L.*	Fruits	+	+	+	+	+
10	*Solanum torvum Sw.*	Fruits	+	+	+	+	+

+ indicates the present

4.4.2. Total Phenolic Content (TPC)

The total phenolic content was measured using Folin-Ciocalteu reagent and the results were derived from the standard calibration curve (y = 0.0045x - 0.0148, R^2 = 0.995) of gallic acid (10- 100μg/mL) and expressed in GAE (gallic acid equivalents)/g (Table 10). The total phenolic content in methanol extract ranged from 9.87 to 29.51 mg/g. *S. anguivi* (29.51 mg/g) had the highest phenolic contents while *S. torvum* (9.87 mg/g) had the lowest content.

4.4.3. Total Flavonoid Content (TFC)

To determine the total flavonoid content aluminium chloride colorimetric method was used and the results was derived from standard calibration curve (y = 0.0006 + 0.0013, R^2 = 0.998), expressed in quercetin equivalents (QE) /g. The total flavonoids content ranged from 8.82 ± 0.002 mg/g to 35.15 ± 0.03 mg/g respectively. *C. annuum* contained the highest flavonoid while *S. betaceum* had the lowest flavonoid content (Table 10).

4.4.4. Total Anthocyanin activity

The total anthocyanin content of methanolic extracts of Solanaceae species varied from 0.069 to 0.91 mg/g (Table 10). *P. angulata* showed the highest and *C. annuum* showed the lowest total anthocyanin content.

Table 10: Quantitative phytochemical analysis of Solanaceae plants.

Sl. No	**Species Name**	**Total Protein Content (mg/g)**	**Total Carbohydrate Content (mg/g)**	**Total Phenolic Content (mg/g)**	**Total Flavonoids Content (mg/g)**	**Total Antho-cyanin content (mg/g)**
1.	*Capsicum annuum* L.	24.75± 0.005	19.12±0.004	20.03± 0.006	35.15± 0.034	0.069
2.	*Capsicum frutescens* L.	22.95± 0.058	24.49± 0.009	19.14± 0.004	32.24± 0.001	0.075
3.	*Solanum betaceum* Cav.	16.93± 0.004	18.19± 0.012	12.30± 0.008	8.82± 0.002	0.45
4.	*Lycopersicon esculentum* Mill.	14.09± 0.004	25.27± 0.041	12.52± 0.007	16.56± 0.001	0.91
5.	*Physalis angulata* L.	17.05± 0.013	35.64± 0.011	21.57± 0.004	30.50± 0.002	0.75
6.	*Solanum americanum* Mill.	19.18± 0.038	16.48± 0.022	16.27± 0.005	23.20± 0.003	0.5
7.	*Solanum anguivi* Lam.	12.04± 0.007	26.95± 0.217	29.51± 0.004	16.56± 0.001	0.38

8.	*Solanum incanum* L.	21.76± 0.055	20.41± 0.011	14.95± 0.008	21.21± 0.002	0.35
9.	*Solanum melongena* L.	28.49± 0.058	18.54± 0.019	15.61± 0.006	19.66± 0.002	0.25
10.	*Solanum torvum* Sw.	6.1± 0.011	15.19± 0.012	9.87± 0.006	11.92± 0.037	0.15

4.5. Determination of nutrient composition

The nutrient composition of edible parts of Solanaceae plants is presented in Tables 10. The protein content of the edible parts ranged from 6.1 mg/g (*S. torvum)* to 28.49 mg/g (*S. melongena).* The carbohydrate content varied from 15.19 mg/g (*S. torvum)* to 35.64 mg/g (*P. angulata).* The mineral compositions found in the study are presented in Table 11. High values of Na, Mg, Ca, and K were found in all the samples and a considerable amount of Fe, Mn, Cu, Zn were also observed. The toxic mineral ions such as Pb and Ni were absent in the studied samples.

Table 11: Element analysis of Solanaceae Plants.

		Element Concentration (mg/kg)									
Sl No.	**Species Name**	**Ca**	**Cu**	**Fe**	**Mn**	**Zn**	**K**	**Mg**	**Na**	**Ni**	**Pb**
1	*Capsicum annuum* L.	1.39	0.019	0.23	0.05	0.072	1.4	0.02	1.4	0	0
2	*Capsicum frutescens* L.	1.68	0.02	0.41	0.05	0.17	0.9	6.56	1.71	0	0
3	*Solanum betaceum* Cav.	1.95	0.021	0.36	0.02	0.077	3.4	5.97	34.3	0	0
4	*Lycopersicon esculentum* Mill.	2.59	0.014	0.39	0.02	0.058	4.2	1	44.7	0	0
5	*Physalis angulata* L.	1.65	0.019	0.27	0.04	0.12	4	5.7	45.1	0	0
6	*Solanum americanum* Mill.	1.89	0.018	0.28	0.1	0.15	3.5	6.54	6.78	0	0

7	*Solanum anguivi* Lam.	2.79	0.02	0.26	0.04	0.12	2	4.05	2.23	0	0
8	*Solanum incanum* L.	2.34	0.039	3.24	0.12	0.11	2.2	1.5	34.1	0	0
9	*Solanum melongena* L.	2.73	0.039	0.43	0.12	0.12	1.3	2.29	9.24	0	0
10	*Solanum torvum* Sw.	5.46	0.045	0.4	0.1	0.1	2.3	1.14	43.4	0	0

4.6. Enzymatic Antioxidant activity

4.6.1. DPPH radical scavenging antioxidant activity

The antioxidant capacity of plant extracts had significant scavenging activities on DPPH that increased with an increase in concentration (10-100μg/ml) as shown in Figure 12. The IC_{50} value was calculated to determine the concentration of the sample required to inhibit 50% of free radicals and a lower IC_{50} value indicated higher antioxidant activity (Li et al. 2009). The present study shows that the free radical scavenging activities of the extracts are concentration-dependent and comparable to ascorbic acid. The IC_{50} of the extracts and ascorbic acid observed in the study are detailed in Table 12. Among the extracts, *L. esculentum* (34 μg/ml) showed the strongest IC_{50} value which was comparable to that of purified ascorbic acid.

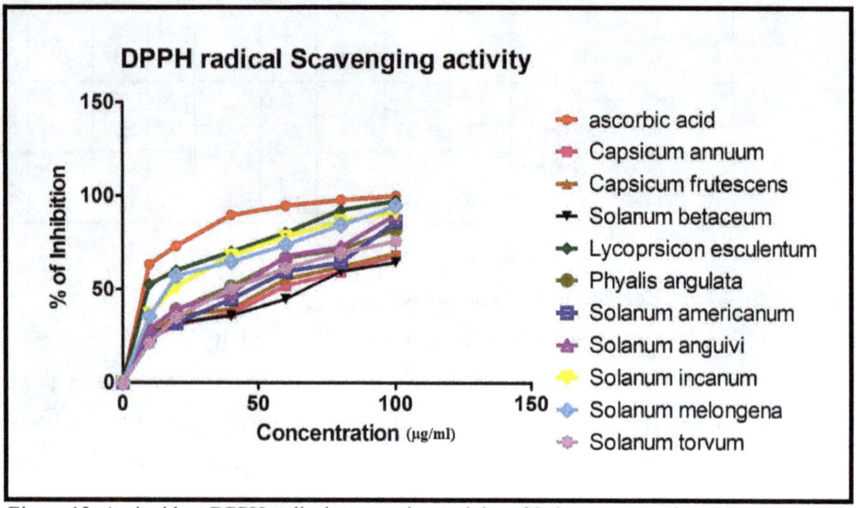

Figure 12: Antioxidant DPPH radical scavenging activity of Solanaceae species.

4.6.2. Total CAT, APX, and SOD activity

The total CAT, APX, and SOD activity of the Solanaceae species are shown in Table 12. H_2O_2 decomposed per minute for catalase activity ranges from 0.32 mM to 6.31 mM (Table 14). *S. anguivi* decomposed 6.31 mM H_2O_2 per minute. *S. melongena* decomposed the least amount of H_2O_2 per minute among the studied samples. H_2O_2 decomposed per minute for APX activity ranges from 0.87 mM to 7.94 mM. *S. betaceum* decomposed 7.94 mM H_2O_2 per minute showing the highest APX activity and *C. frutescens.* (0.87 mM H_2O_2 per minute) showed the lowest APX activity. The SOD activity ranged from 0.23U to 1.63U. *L. esculentum* showed the highest SOD enzymatic activity while *S. americanum* possessed the lowest SOD enzymatic activity.

Table 12: Enzymatic antioxidant activity of Solanaceae plants.

Species Name	Catalase (mM H2O2 decomposed/min	APX (mM H2O2 decompoed/min)	SOD (Unit)	Total Antioxidant IC50 (µg/ml)
Capsicum annuum L.	4.21	0.93	1.46	49.51
Capsicum frutescens L.	3.72	0.87	1.38	45.72
Solanum betaceum Cav.	4.2	7.94	1.32	54
Lycopersicon esculentum Mill.	0.89	0.98	1.63	34
Physalis angulata L.	0.52	1.24	0.94	41.81
Solanum americanum Mill.	5.61	1.4	0.23	53.4
Solanum anguivi Lam.	6.31	3.26	1.3	49.46
Solanum incanum L.	3.73	2.8	0.56	44.24
Solanum melongena L.	0.32	3.27	0.84	35.67
Solanum torvum Sw.	5.89	4.29	0.29	36.09
Ascorbic acid				29.23

4.7. Antibacterial activity

The microbial growth inhibition of the methanolic extracts (Figure 13) is summarized in Table 13. The antibacterial activities of extracts show strong effective inhibition activity against *Escherichia coli, Bacillus subtilis,* and *Pseudomonas aeruginosa.* The maximum antibacterial activity was shown by *C. annuum* and the least by *S. torvum.*

Table 13: Antibacterial Activity of Solanaceae plant extracts

Species Name	*Escherichia coli Inhibition Zone (mm)*			*Bacillus subtilis Inhibition Zone (mm)*			*Pseudmonas areuginosa Inhibition Zone (mm)*		
	20 mg/ml	40 mg/ml	60 mg/ml	20 mg/ml	40 mg/ml	60 mg/ml	20 mg/ml	40 mg/ml	60 mg/ml
Streptomycin (Positive control)	15.44 ± 0.58			20.44 ± 0.78			22.73 ± 0.69		
Capsicum annuum L.	7.33 ± 0.33	9.97 ± 0.54	11.9 ± 0.58	10.67 ± 0.34	14.67 ± 0.89	16.67 ± 1.76	9.33 ± 1.21	13 ± 0.57	16.34 ± 0.67
Capsicum frutescens L.	6.12 ± 0.46	8.2 ± 0.74	10 ± 1.31	9.45 ± 0.72	15.2 ± 0.42	16.62 ± 1.19	8.21 ± 0.67	11. 06 ± 1.12	15 ± 0.21
Solanum betaceum	3.45 ± 0.39	7.97 ± 0.89	10.89 ± 0.42	5.67 ± 0.38	8.88 ± 1.12	11.43 ± 0.47	6.33 ± 0.19	9.78 ± 0.49	12.22 ± 0.29
Lycopersicon esculentum Mill.	9.01 ± 0.54	11.74 ± 0.23	13 ± 0.36	8.5 ± 1.07	12.1 ± 0.22	14.2 ± 0.44	6.4 ± 0.56	9.2 ± 0.67	14 ± 0.78
Physalis angulata L.	5.03 ± 1.42	7.41 ± 0.44	10.2 ± 0.22	4.07 ± 0.33	6.51 ± 0.81	9.5 ± 0.61	5.6 ± 0.88	7.8 ± 1.41	10.7 ± 0.74
Solanum americanum Mill.	5.76 ± 1.12	7.45 ± 0.67	9.01 ± 1.21	7.22 ± 0.11	10.67 ± 0.96	12.45 ± 0.11	8.56 ± 0.68	11.33 ± 0.33	12.22 ± 0.29
Solanum anguivi Lam.	6.33 ± 0.33	8.22 ± 0.11	10.66 ± 0.48	7.66 ± 0.11	11.33 ± 0.44	14.55 ± 0.58	7.44 ± 0.56	9.27 ± 0.63	11.67 ± 1.02
Solanum incanum L.	4.27 ± 0.11	7.01 ± 0.41	7.92 ± 0.34	4.22 ± 0.64	8 ± 0.49	9.3 ± 0.24	5.1 ± 0.97	7.2 ± 0.87	8.9 ± 0.52
Solanum melongena L.	5.89 ± 0.22	8.43 ± 0.29	10.66 ± 0.19	6.22 ± 0.22	9.77 ± 0.39	12.33 ± 0.19	5.1 ± 0.55	6.67 ± 0.38	11.17 ± 0.19
Solanum torvum Sw.	4.44 ± 0.56	6.27 ± 0.63	7.44 ± 0.29	3.5 ± 0.40	5.76 ± 0.34	6.9 ± 1.45	3.9 ± 0.89	6.2 ± 1.12	7.1 ± 0.92

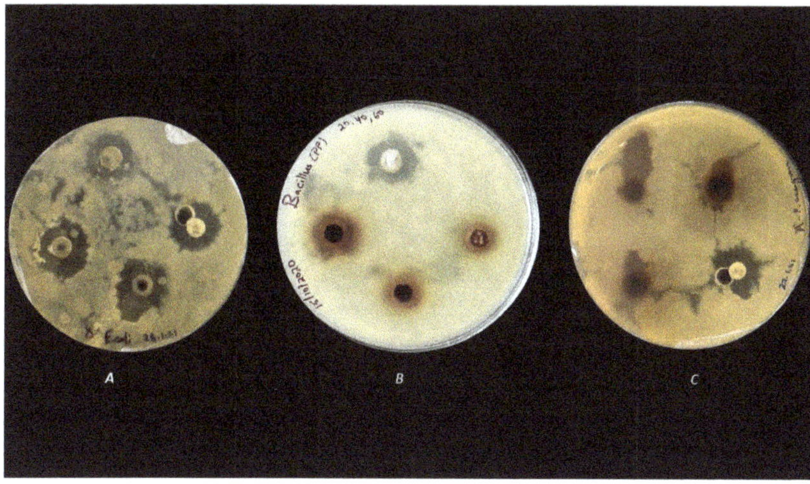

Figure 13: Antimicrobial zone of inhibition of plant extract (*S. anguivi*) against tested bacterial strains. A – *B. subtilis* B – *E. coli*; C – *P. aeruginosa.*

4.8. Correlation between Antioxidant DPPH Scavenging Activity, Total phenolic, and Flavonoid Content

A correlation analysis was performed for total phenol, flavonoid content against antioxidant activities detected in Solanaceae plants species (Figure 14). A significant correlation was observed between total phenol, total flavonoid content and antioxidant potential ($y = 0.515x$, $R^2 = 0.73$ and $y = 0.411x$, $R^2 = 0.68$, $p \leq 0.05$ respectively). The strong correlation means that the phenolic and flavonoid contents contributed significantly to the antioxidant activity (Margaryan et al. 2017). The present analysis indicates that the antioxidant activity of the studied samples is strongly correlated with the high content of total phenolic and flavonoid that can play as reductones by donating electrons and reacting with free radicals thereby converting into more stable products (Margaryan et al. 2017).

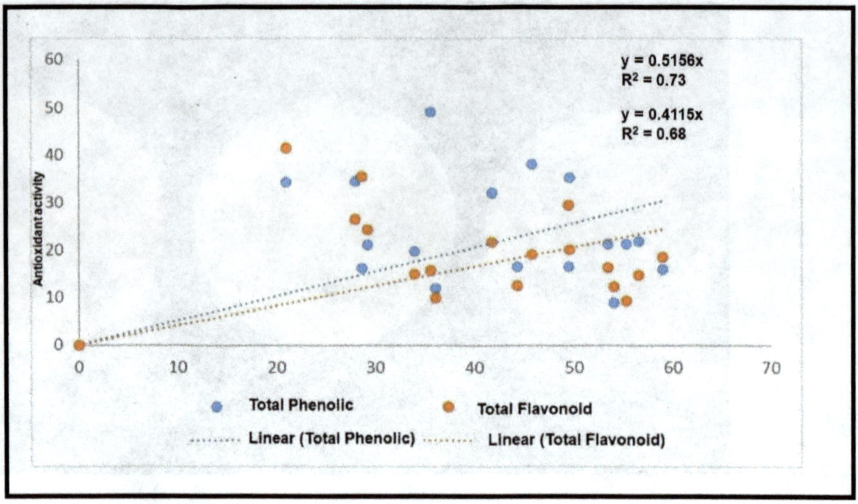

Figure 14: Correlation analysis between Antioxidant activity, Total Phenolic, and Flavonoid Content.

4.9. FT-IR Analysis

The FT-IR spectra and the functional group of compounds present in the edible parts of Solanaceae plants species are shown in Figures 15 a & b. The FT-IR analysis (Table 14) showed the presence of alkaloids (N-H stretch), polyphenols and flavonoids (O-H stretching), terpenes (C-H stretching). Table 14 shows the presence of functional groups such as aldehydes, alkenes, amines, amides, alcohols, phenols, aromatics, carboxylic acids, esters, ethers, organic halogen compounds in the studied samples.

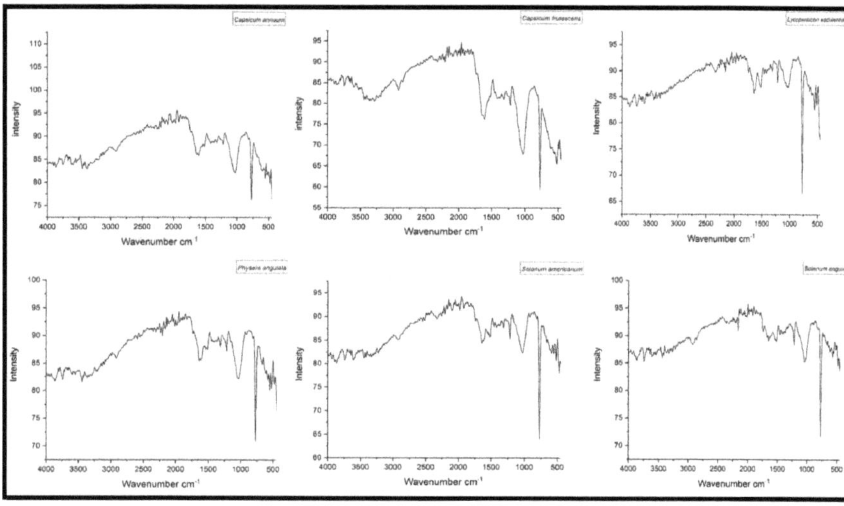

Figure 15a: FT-IR spectra of Solanaceae plants.

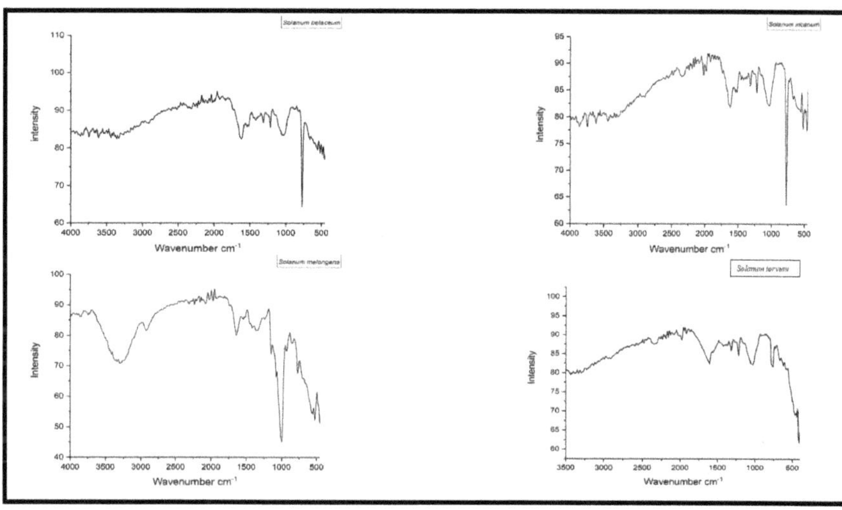

Figure 15b: FT-IR spectra of Solanaceae plants.

Table 14: Evaluation of FT-IR spectrum of Solanaceae plants.

Frequency range (cm-1)	Peak wavenumber (cm-1)										Functional group
3870 - 3550	*C. annuum*	*C. frutescens*	*L. esculentum*	*P. angulata*	*S. Americanum*	*S. anguivi*	*S. betaceum*	*S. incanum*	*S. melongena*	*S. torvum*	O-H stretch alcohol
3500 - 3200	3750	3672	3672.5	3865	3742	3865	3834	3741.9		3672	O-H stretch vibration presence of alcohols, phenols
3300 - 2850	3441	3441	3387	3441	3449	3649	3487	3417.9	3364	3325	O-H stretch vibration, carboxylic acids
2500 - 2300		2916		2924	2924	3225		3302.1	2924		C-H stretch vibration, alkenes
2260 - 2100			2330	2307			2446			2484	C=C stretch vibration, alkynes
1990 - 1739		2160	2152.6	2137	2207	2160		2237.4		2237	Ester C=O stretch, lipid, triglycerides
1700 - 1600		1836	1743.7		1983	1921	1844	1975.1		1975	C=C stretch vibration, alkenes
1550 - 1475		1605			1643	1651			1620		N-O asymmetric stretch, nitro compounds
1470 - 1400					1520			1543.1			C-C stretch vibration, aromatics
1400 - 1320			1458.2		1458	1458	1420			1458	N-O stretch vibration, nitro compounds
1300 - 1290		1319		1319					1319	1319	C-O stretch vibration, alcohol, carboxylic acids, esters, ether

1275 - 1150											C-H wag stretch vibration, alkyl halides
1020 - 1000		1219	1219	1219	1219	1219		1219	1242	1219	C-N stretch vibration, aliphatic amines
990 - 800		1026			1034	1034			1034	1034	N-H wag stretch vibration, primary & secondary amines
790 - 690											C(triple bond)C-HC-H bend stretch vibration, alkynes
680 - 510	772	772		771.5			77.5	741.53	779	772	C-Br stretch vibration, alkyl halides, glycogen
490 - 400	556	517	640.37	671.2	594.1	617	664	616.92	556	617	Halogen compound

(Schwanninger et al. 2004; Bhat 2011

Chapter 5

Discussion

People of Mizoram receive their knowledge on traditional medicines from their ancestors orally. Proper documentation of this knowledge is the need of the hour. My present study collected and identified 20 species of Solanaceae that were well distributed and commonly found in Mizoram. Information on various genera, species, habit-wise differentiation, botanical name, common name, local name, species description, and ethnobotanical uses were documented. Among the documented plants, the genus *Solanum* (45%) was the dominant one followed by *Capsicum* (10%), and *Physalis* (10%). One genus each of *Brugmansia, Datura, Lycianthes, Lycopersicon, Nicotiana, Petunia* was found from the study sites. The specimens were mostly found in forests, grazing land, farms, roadsides, and plantation areas. Different plant parts were found to be useful for the treatment of various diseases. The most commonly used parts for preparations of traditional medicines were fruits, leaves, and whole plants in which decoction, powdered form, and boiling with water were the common methods of preparations. It was observed that most of the Solanaceae plants were shrubs and herbs. Because leaves of medicinal plant species were reported to be harvested for most remedy preparations, gathering leaves could be promoted as a more sustainable method because, in most cases, at least several leaves are leftover on the parent plant allowing it to carry on its life functions (Lulekal et al. 2013). The plant parts or extracts were either administered orally, applied externally, as cream. Taking the herbal mixtures orally routinely treated internal illness, whereas crushing and applying the therapeutic plant portion to the infected area treated tooth infections. Creaming the herbal mixture on the infected area treated skin infections. To estimate and fix the amount or dosage of herbal medicine, local people traditionally used different units of measurement such as finger length for measuring root, bark, etc., a pinch for powdered form, and numbers for measuring the leaves, seeds, and fruits. The preparations are taken with known dosages that are generally measured with a spoon, and other tools, and the majority of medicinal plants recommended and administered to patients are administered in non-standardized doses. Herbal remedies were prescribed to the patients differently according to their age groups, the severity of the illness, and other conditions. For example, the prescription dose for children was much lower than that of adults. The ethnomedicinal knowledge is greater among the elders and the traditional knowledge is mostly transferred to their family members verbally. Here, it was found that transmission of indigenous knowledge to the next generation was inadequate, which could lead to the practice's extinction. On the other hand, modernization has given an impact on the transmission of medicinal knowledge to the next generation. This might be due to fading of interest among the younger generation in indigenous knowledge and the abundance of easily accessible synthetic drugs. However, many plants species of Solanaceae are important sources for treating human ailments. My study could document the Solanaceae plants used in Mizoram by traditional healers, local practitioners for the treatment of various ailments. The study also suggests that further investigation of these plants can provide a better understanding of their roles in the treatment of various ailments.

The primer universality is a critical condition for a functional DNA barcode (Kress et al. 2005). In my study, two barcode loci (*rbcL* and *ITS2)* gave the best performances in PCR amplification and sequencing of the Solanaceae. The low success rate of *matk* and *rpoC1* could be due to difficulty with secondary structure development that had resulted in low-quality sequence data, multiple copies, and other issues (Starr et al. 2009; Xiao et al. 2010). As a result, these regions are unlikely to be effective as a universal barcode for Solanaceae, however, it may be useful in other circumstances. The morphological characters such as perianth, inflorescence type, petals size, and arrangement, etc. have been used to differentiate the species of Solanaceae. Several polymorphic morphological traits were useful for characterization at the genus level, however, most species had inconspicuous flowers that were difficult to find or collect, generic delimitation within the family was difficult and must be identified using vegetative characters, so, misidentifications were unavoidable. Due to scant information about the species, errors of identification based on morphological characters sometimes occurs. Other factors like environment and geographical conditions also added to errors in the identification and therefore, hampered accurate identification of Solanaceae. Other workers of Solanaceae had also encountered similar problems (Ganaie et al. 2018). Hence, DNA barcoding can be a suitable candidate for detecting errors in species identification (Gonzalez et al. 2009). To solve identification errors based only on morphological characters, tree-based and sequence similarity-based techniques using DNA barcoding in combination with morphology are highly recommended (Huang et al. 2015). The plant working Group of the Consortium (CBOL group 2009) recommended two regions combination of *rbcL* and *matK* for plant DNA barcoding. But *matK* is one of the most rapidly evolving plastid regions and needed to improve in many ways such as primer universality, the efficiency of PCR amplification, etc. (Chase et al. 2007; Hollingsworth et al. 2016). Other combinations of barcodes such as *psbA-trnH, Ycf5, rp0C1* were also attempted for Solanaceae plants, however, due to low sequencing quality, low discrimination potentials, those primers were not suggested for barcoding (Thomas 2009). Then, later studies reported *ITS2* and *rbcL* as one of the best universal primer pairs for plant identification (Chen et al. 2010). Solanaceae necessitates a close examination of the traits used to define specific and generic terms. Solanaceae, in particular, can benefit from precise recognition using DNA barcoding because it is the most diverse and most commonly used plant in Mizoram. And, due to their high species identification and discrimination ability, *ITS2* and *rbcL* were proposed as important DNA barcodes for plants (Kress et al. 2005; Li et al. 2011), and in this study, they did provide ideal species identification. Here, *ITS2* and *rbcL* regions showed high efficiency of PCR amplification and high sequence quality at the same time they had high species discrimination potential. Earlier, it was also already suggested that *ITS2* and *rbcL* were suitable markers for taxonomic classification and phylogenetic reconstruction in eukaryotes (Schultz et al. 2005; Coleman 2007; Miao et al. 2008).Therefore, *ITS2* and *rbcL* can be regarded as ideal candidates for the barcoding of Solanaceae species. Consequently, Neighbor-joining (NJ) analyses showed significant phylogenetic resolution for Solanaceae at both generic and intrageneric levels. The analyses showed that *Ipomea batatas*, which was used as an outgroup, was discriminated from the Solanaceae plants and grouped into separate clade while members of the same genus were mostly grouped under the same clade. Now, it can be concluded that DNA barcoding is a useful technique for resolving phylogenetic relationships at generic and species levels. Solanaceae is one of the most economically important families among the angiosperms and their commercial values have gained more attention. Therefore, a more accurate, reliable, and simple method of identification was highly essential.

The preliminary qualitative phytochemical analysis of edible plants of Solanaceae of Mizoram revealed the presence of various bioactive compounds that were reported to have different biological and therapeutic properties. Alkaloids are nitrogenous compounds having antioxidant potential and have been used in folk medicine (Quezada et al. 2006). Saponin is commonly used as a natural antioxidant and also promotes apoptosis in tumor cells (Podolak et al. 2010; Bi et al. 2012). Tannins are well-known antimicrobial agents (Sodipo et al. 1991), have antioxidant potential, and have been used as active ingredients in medicine and beverages (Amarowicz and Troszynka 2003). Likewise, flavonoids have antioxidant properties and can prevent cell damage, providing anticancer and anti-inflammatory activities (Salah et al. 1995; Okwu 2004). Similarly, it has been reported that the presence of terpenoids influences antimicrobial properties (Mazher et al. 2016), and has been used as a protective agent against oxidative stress-induced diseases (Grassmann 2005).

Plants are a diverse source of phenolic compounds with different functions and the majority are bioactive compounds with anti-cancer, anti-viral, antioxidant, and anti-bacterial potentials (Manach et al. 2004). The total amount of phenol found in the extracts was comparable to previous reports by Elekofehinti et al. (2013), Oyeyemi et al. (2015), Yousaf et al. (2013). Among the extracts, *S. anguivi* had the highest amount of phenol and this might be the reason that the plant had been often used for the treatment of various skin diseases. Flavonoids are bioactive compounds belonging to the polyphenolic class and constitute the major antioxidant in fruits, plants and have advantageous effects on human health. Due to their high antioxidant properties, flavonoids are important sources of the human diet (Calado et al. 2015). They have high potential in antimicrobial, anticancer, anti-inflammatory, and anti-allergic activities due to their ability to scavenge reactive oxygen species (ROS) consisting of free radicals (Montoro et al. 2005). In my study, the total flavonoid obtained was slightly higher than previous reports (Hassan and Bakar 2013; Mutalib et al. 2017;

Vasco et al. 2009). Even a positive correlation of flavonoid and phenol content with a high antioxidant potential of the extracts was also recognized (Table 4). Thus, the extracts, filled with high phenol and flavonoids, could be good sources of antioxidants thereby lowering the risk of diseases triggered by oxidative stress and also improving overall antioxidant capacity. Anthocyanins are involved in enzymatic reactions in the flavonoid biosynthesis pathway (Li et al. 2019). Anthocyanins also provide protection against certain chronic diseases such as hyperglycemia (Tsuda et al. 2003), inhibit the growth of tumor cells in humans (Wang et al. 2017; Zhao et al. 2005), and also improve vision (Kalt et al. 2014). Anthocyanins have high antioxidant potential, antibacterial properties and are also used as natural food colorants (Naz et al. 2007). In my study, the highest total anthocyanin content was found in *L. esculentum* (0.91 mg/g) followed by *P. angulata* (0.75 mg/g). *C. annuum* showed the lowest total anthocyanin content (Table 3). TAC was found to be higher than previous reports from *S. nigrum, S. tuberosum, S. lycopersicon, S. melongena, N. tabacum, P. hybrida,* and *Withania somnifera* extracts (Kanungo et al. 2013; Wang et al. 2017). A recent study has also suggested that Solanaceae plants are promising resources for anthocyanin extraction (Li et al. 2019). The demand for anthocyanins is increasing in commercial industries and pharmaceuticals for the treatment of various diseases and also in beverage industries (Zhang et al. 2003). So, Solanaceae plants could be good sources of anthocyanins for various pharmaceutical and other commercial industries.

Antioxidants present in food are gaining prominence due to their significant function in maintaining human health by preventing diseases by inhibiting free radicals that are responsible for the spread of various diseases such as cancer, neurodegenerative disorders, etc. The IC_{50} for DPPH of *L. esculentum* was found to be the lowest among the studied plants indicating strong antioxidant potential while *S. torvum* showed the highest DPPH. The phenol and flavonoids are multifunctional bioactive compounds that are antioxidant, antimicrobial, anti-inflammatory, and anti-cancer agents. Several studies concluded that these multifunctional bioactive compounds are the major contributors to the antioxidant potential of plant extracts (Shahidi and Ambigaipalan 2015). Hence, the free radical scavenging capacity observed in our study could be due to high levels of phenol and flavonoid in the extracts. This is in agreement with a report, showing higher free radical scavenging activity with higher overall phenolic and flavonoid content (Zhang et al. 2016). Hence, the present study reveals that *L. esculentum* has a strong antioxidant potential. This property may be due to higher phenol, flavonoid, and anthocyanin content, which are required for scavenging activity, in *L. esculentum*. It is also known that the amount of phenolic and flavonoid content in plants is responsible for the free radical scavenging activity. My study suggests that the extracts of edible plants of Solanaceae display high antioxidant capacities. Environmental conditions like extreme temperature, water stress, high light intensity can cause oxidative damage by over-production of toxic ROS (Bowler et al. 1992). However, plants can protect themselves against oxidative damage by an antioxidant system such as anti-oxidative enzymes and non-enzymatic compounds (Mittler 2002). Plants contain various anti-oxidative enzymes including SOD, CAT, APX, etc. (Wang et al. 2009b). SOD converts superoxide radicals into hydrogen peroxide, APX uses ascorbate as an electron donor to reduce hydrogen peroxide to water, CAT dismutates hydrogen peroxide into water and oxygen (Wang et al. 2009b). Living organisms can protect themselves from the toxic effects of ROS. SOD, APX, and CAT are enzymes that help in detoxifying ROS. Increased levels of SOD, APX, and CAT can lead to enhanced oxidative stress protection (Gupta et al. 1993). Few previous reports have also shown that Solanaceae plants have potential activities of SOD, APX, and CAT (Yu et al. 1998; Tang et al. 2006; Kanungo et al. 2013). My investigation also indicated that the Solanaceae plants could be good sources of SOD, APX, and CAT that could have significant value in reducing stress oxidative reaction. Owing to their high antioxidant capacity, these plants can serve as good sources of antioxidants for pharmaceutical and nutraceutical formulations.

It was also found that the carbohydrate content was much higher than a previous report by Akoto et al. (2015) in *S. torvum* (7.033 mg/g). The protein content in the plant extracts was also found to be higher than a previously reported value of 2.32 mg/g of the plant extract (Agoreyo et al. 2012). High values of protein and carbohydrate indicate rich in essential nutrients that could be utilized for enhancing nutrition. Then, the mineral ion compositions of the plants were also relatively high in all the studied samples. Dietary intake of potassium has a significant effect on coronary heart diseases by reducing blood pressure (Weaver 2013). Calcium is an essential mineral ion for the human diet and is involved in cell differentiation, muscle, and bone formation (Roberts et al. 2000). Sodium is required for physiological processes, body fluid balance, and cellular homeostasis (Abdulrahman 2004). Magnesium is essential for the circulatory system and is important for various metabolisms (Nwauzoma and Dawari 2013). My study also showed the presence of micronutrients such as Fe, Cu, Mn, Zn. These micronutrients are required for metabolic processes like respiration and DNA synthesis (Lieu et al. 2001). Thus, my findings suggested the effective utilization of these plants as a source of minerals and nutrient supplements.

The global burden of infectious diseases caused by bacterial organisms poses a severe public health risk (Eggleston et al. 2010). Antibiotic treatment is the recommended method for treating bacterial infections, however, the evolution of antimicrobial resistance and toxicity concerns limit the use of antibacterial medicines (Zhang et al. 2006; Malini et al. 2013). Antibiotic safety and efficacy constraints supplement biological research on the antibacterial role of plants due to comparable toxicity and efficacy (Alviano and Alviano 2009). Antibiotic resistance is an epidemic that continues to plague the healthcare system in both developing and developed countries around the world. The appearance and dissemination of multidrug-resistant pathogens have significantly jeopardized conventional antibacterial therapy. This has led to a hunt for new antimicrobial sources preferably from plants that contain various bioactive compounds with established therapeutic properties. A study of the zone of inhibition against bacterial strains with Solanaceae plant extracts revealed that all bacteria subjected to methanol extract showed zone of inhibitions. The present study was undertaken to assess the antimicrobial efficacy of edible plants of Solanaceae against multi-resistant bacterial strains- *B. subtilis, E. coli,* and *P. aeruginosa*. Results indicated that the plant extracts exhibited significant antibacterial activities against the tested bacterial isolates. *L. esculentum* extract showed maximum activity against all three pathogens. The inhibition was even higher than one reported on methanol extracts of other Solanaceae plants (Rawani et al. 2013). One of the most serious challenges to humanity is the rise of multidrug resistance by pathogens. The application of effective plant extracts might be a valuable option in combating this phenomenon. Hence, the plants under current investigation could be useful in combating antidrug resistance in the tested bacterial strains. However, further investigations are sought to evaluate anti-viral, anti-fungal, and anti-parasitic activities to harness the potentials of the plants.

Another important aspect of my study was to identify the functional groups found in the plant extracts using FTIR. This analysis helps in the identification of chemical composition, elucidation of the chemical structure and to understand the importance of functional groups as bioactive compounds for phytopharmaceuticals formulations. The plants have shown similar infra-red spectrum and some intense bands at various frequencies which define the presence of O-H (hydroxyl), O-H stretch (carboxylic acid), O-H bend (phenol or tertiary alcohol) C-H stretch (alkanes), C=C-C (aromatic compounds), C=C stretch (ketone), N-O (nitro compound), C-O (ether), C-N (aromatic primary amines), N-H(amines), C≡C (carbonyl), C-Br (aliphatic Bromo compounds) (Table 7) groups. The presence of these functional groups indicates the presence of different metabolites such as aldehydes, alkanes, alkenes, alkynes, alkyl halides, aliphatic amines, primary and secondary amines, alcohols, aromatics, carboxylic acids, esters, ethers, glycogen, hydroxyl, lipid, organic halogen compounds, nitro compounds, phenols, and triglycerides, that are integral parts of most of the secondary metabolites such as alkaloids, flavonoids, tannins, terpenoids and polyphenol (Poojary et al. 2015). Functional groups in the plants can be used in different pharmaceutical products such as for anti-cancers, anti-ulcers, jaundice, headache, stomach ache, and anti-inflammatory drugs; or as sources of antimicrobial, antioxidant compounds, etc. (Baker 1982; Skoog et al. 2007; Maobe and Nyarango 2013). This may also be the reason why traditionally these plants are used by the locals in the treatment of stomach aches, as anti-inflammatory medicine, etc. (Table 1). The phytochemical screening and FTIR analysis showed that various bioactive compounds were found in the plant extracts that can be used as an active antioxidant and anti-microbial agents of plant origin. The current study also revealed clear discrimination between the plant parts tested (leaf, fruit whole plant, etc.), displaying significant heterogeneity for the identification of bioactive phytochemicals that can be used as herbal medicines. However, further studies are necessary to evaluate *in vivo* biological activities of the bioactive phytochemicals for designing effective phytopharmaceutical formulations.

Chapter 6

Conclusion

A study on ethnobotanical use of Solanaceae plants in Mizoram, India indicated that local people for their primary healthcare system have extensively used the Solanaceae plants. Twenty Solanaceae species were documented where shrubs were found to be the dominant growth form for the preparation of traditional medicine. Leaves were found to be the most common plant part used for traditional medicine preparations. Despite its high floristic diversity, Mizoram remains under-explored from an ethnobotanical standpoint. The indigenous knowledge of ethnobotanical uses is culturally significant and can be used to inform future research and promote ethnopharmacological advancements. Hence, the present study of plant knowledge can contribute to a better understanding of the factors that influence changes in plant usage and perception in various sociocultural contexts. The *ITS2* and *rbcL* barcode loci used in the present study showed a positive result in species identification and reconstructing the phylogenetic relationship of Solanaceae. Since, DNA barcoding is useful for the conservation of rare species and the prosecution of forest crimes, the present study recommends using DNA barcodes in conjunction with other methodologies to build an appropriate management strategy for the long-term protection of Solanaceae. And a combination of *ITS2* and *rbcL* DNA barcodes could prove significant for the evaluation of biodiversity and discovery where adoption of a single barcode might drastically cut the time and expenses associated with species identification. However, the study highlights the crucial need for further data from more taxa and sequence regions to help resolve challenges in Solanaceae taxonomy and conservation. Bioactive compound compositions, antioxidant activities, nutrient compositions, and antimicrobial potential of edible plants of Solanaceae from Mizoram, India were also analyzed. These plants contain various bioactive phytochemicals, antimicrobial agents with various functional groups and have promising nutritional and antioxidant potential. Results demonstrated that these plants could be used as an easily accessible source of natural bioactive compounds with antioxidant and antimicrobial potentials and can also substitute synthetic drugs. The present investigation using FTIR analysis showed the presence of various functional groups such as carbohydrates, glycogen, etc. As a result, the FTIR spectrum, which objectively depicts the landscape of bioactive constituents in a complex system, is the most reliable approach for validating and identifying mix-substances in traditional medicine and herbal medicine. Further research could aid in the discovery of new bioactive compounds in these therapeutic plants. To the best of my understanding, this is the first report of Solanaceae plants from Mizoram that investigates ethnobotanical uses, identification using DNA barcodes, identification of bioactive compounds, mineral nutrient contents, antimicrobial potential, determination of antioxidant, and functional groups. Further studies on these plant species could open a new perspective for developing novel health-promoting agents in the pharmaceutical and nutraceutical industries.

References

Abdel-Aal ESM, Hucl P (1999) A rapid method for quantifying total anthocyanins in blue aleurone and purple pericarp wheats. *Cereal Chemistry* 76(3):350-354.

Abdulrahman FI (2004) Studies in the Chemical Contents and Pharmacological Activities of the root-bark extract of *Vitex doniana* (Black Plum). PhD Thesis, University of Maiduguri, Maiduguri, 55-82.

Agoreyo BO, Obansa ES, Obanor EO (2012) Comparative nutritional and phytochemical analyses of two varieties of *Solanum melongena. Science World Journal.* 7(1): 1597-6343.

Akoto O, Borquaye LS, Howard AS, Konwuruk N (2015) Nutritional and mineral composition of the fruits of *Solanum torvum* from Ghana. *International Journal of Biological Chemistry* 1 (4): 222-226.

Ali MA, Gyulai G, Hidvegi N, Kerti B, Al Hemaid FMA, Pandey AK, Lee J (2014) The changing epitome of species identification – DNA barcoding. *Saudi Journal of Biological Science* 21:204-231.

Almoulah NF (2017) Phytochemical study and biological evaluation of five plants belonging to the family Solanaceae. Chemical Enineering. *Universite de Lorraine* 2017LORR0184.

Alviano DS, Alviano CS (2009) Plant extracts: search for new alternatives to treat microbial diseases. *Current Pharmaceutical Biotechnology* 10(1):106-121.

Amarowicz A, Troszynska A (2003) Antioxidant and antiradical activity of extract of pea and its fractions of low molecular phenolics and tannins. *Polish. Journal of Food and Nutrition Sciences* 12: 10–15.

Anderson M, Cornish E, Mau SL et al. (1986) Cloning of cDNA for a stylar glycoprotein associated with expression of self-incompatibility in *Nicotiana alata. Nature* 321, 38–44. https://doi.org/10.1038/321038a0.

Andersson DI, Hughes D (2010) Antibiotic resistance and its cost: is it possible to reverse resistance? *Nature Reviews Microbiology* 8:260-271.

Antia BS, Akpan EJ, Okon PA, Umoren IU (2006) Nutritive and anti-nutritive Evaluation of sweet potatoes (*Ipomea batatas*) leaves. *Pakistan Journal of Nutrition* 5166-168.

Ara MT, Rahman AHM, Khokon EH (2011) Taxonomic studies on the family Solanaceae in the Rajshahi University Campus. *Journal of Biodiversity and Environmental Sciences* 1(4):29-34.

Arnous A, Makris DP, Kefalas P (2001) Effect of principal polyphenolic components in relation to antioxidant characteristics of aged red wines. *Journal of Agricultural and Food Chemistry* 49:5736-5742.

Ashfaq M, Asif M, Anjum ZI, Zafar Y (2013) Evaluating the capacity of plant DNA barcodes to discriminate species of cotton (*Gossypium*: Malvaceae). *Molecular Ecology Resources* 13: 574–582.

Baker EA (1982) Chemistry and morphology of plant epicuticular waxes. In: Cutler DF, Alvin KL, Price CE. Eds. The Plant Cuticle, London: Academic Press, 139-165.

Basak M, Sonar AS, Chowdhury M (2017) Phenetic Study of Some Indian Solanaceae based on Foliar Micromorphology, Palynology and Reproductive Potentiality. *Plant Archives* 12(2): 1267-1273.

Beadle GW, Tatum EL (1941) Genetic control of biochemical reactions in Neurospora. *Proceedings of the National Academy of Sciences* 27(11): 499-506.

Bernatzky R, Tanksley SD (1986) Toward a saturated linkage map in tomato based on isozyme and random cDNA sequences. *Genetics* 112(4): 887-98.

Bhat R (2011) Potential use of Fourier Transform Infrared Spectroscopy for identification of molds capable of producing mycotoxins. *International Journal of Food Properties* http://dx.doi.org/ 10.1080/10942912.2011.609629.

Bi L, Tian X, Dou F, Hong L, Tang H, Wang S (2012) New antioxidant and antiglycation active triterpenoid saponins from the root bark of *Aralia taibaiensis*. *Fitoterapia* 83: 234–240. doi: 10.1016/j.fitote.2011.11.002.

Bindler G, Hoeven R, Gunduz I, Plieske J, Ganal M, Rossi L, Gadani F, Donini P (2007) A microsatellite marker based linkage map of tobacco. *Theoretical Applied Genetics* 114(2):341-349.

Bolger A, Scossa F, Bolger M et al. (2014) The genome of the stress-tolerant wild tomato species *Solanum pennellii*. *Natural Genetics* 46: 1034-1038.

Boligon AA, Pereira RP, Feltrin AC, Machado MM, Janovik V, Rocha JBT, Athayde ML (2009) Antioxidant activities of flavonol derivates from the leaves and stem bark of Scutia buxiflia Reiss. *Bioresource Technology* 100:6592-6598.

Bolson M, Smidt EC, Brotto ML, Silva-Pereira V (2015) *ITS* and *trnH-psbA* as efficient DNA barcodes to identify threatened commercial woody angiosperms from Southern BrazilianAtlantic rain forests. *Plos ONE* 10: e0143049.

Bombarely A, Rosli HG, Vrebalov J, Moffett P, Mueller LA, Martin GB (2012) A draft genome sequence of *Nicotiana benthamiana* to enhance molecular plant-microbe biology research. *Molecular Plant-Microbe Interactions* 25(12):1523-30. doi: 10.1094/MPMI-06-12-0148-TA. PMID: 22876960.

Bonierbale MW, Plaisted RL, Tanksley SD (1988) Reveal modes of chromosomal evolution in potato and tomato. *Genetics* 120(4): 1095-1103.

Bostijn N, Hellings M, Van Der Veen M, Vervaet C, De Beer T (2018) In-line UV spectroscopy for the quantification of low-dose active ingredients during the manufacturing of pharmaceutical semi-solid and liquid formulations. *Analytical Chimica Acta* 12: 54–62. doi: 10.1016/j.aca.2018.02.007.

Bowler C, Van-Montagu M, Inze D (1992) Superoxide dismutase and stress tolerance. *Annual Review of Plant Physiology and Plant Molecular Biology* 43: 83-116.

Brangule A, Sukele R, Bandere D (2020) Herbal Medicine characterization perspective using advanced FTIR sample technique – Diffuse Reflectance (DRIFT) and Phtotoacoustic Spectroscopy (PAS). *Frontier in Life Sciences* https://doi.org/10.3389/fpls.2020.00356.

Bunaciu AA, Aboul-Enein HY, Fleschin S (2011) Recent applications of Fourier transform infrared spectrophotometry in herbal medicine analysis. *Applied Spectroscopy Reviews* 46(3):251-260.

Bunaciu AA, Aboul-Enein HY, Fleschin S (2010) Application of Flourier transform infrared spectrophotometry in herbal medicine analysis. *Applied Spectroscopy Reviews* 45(3):206-219.

Bye R, Sosa V (2013) Molecular Phylogeny of the Jimsonweed Genus *Datura (*Solanaceae). *Systematic Botany* 38:818-829.

Calado JCP, Albertao PA, de Oliveira EA, Sisto MH, Frankland ACH, Marcucci MC (2015) Flavonoid contents and antioxidant activity in fruit, vegetables and other types of food. *Agricultural Sciences* 6: 426–435. doi: 10.4236/as.2015. 64042.

Caprar M, Copac CM, Chende DM, Sicora O, Sumalan R, Sicora C (2017) Evaluation of genetic diversity by DNA barcoding of local tomato populations from North-Western Romania. *Notulae Botanicae Horti Agrobotanici* 45:276-279.

CBOL Plant Working Group (2009) A DNA barcode for land plants. *Proceedings of the National Academy of Sciences USA* 106:12794-12797.

Chang CC, Yang M, Wen HM, Chern JC (2002) Estimation of total flavonoid content in propolis by two complementary calorimetric methods. *Journal of Food and Drugs Analysis* 10:178-182.

Chase MW, Cowan RS, Hollingsworth PM, Van den Berg C, Madrinan S, Petersen G, Seberg O, Jorgsensen T, Cameron KM, Carine M, Pedersen N, Hedderson TAJ, Conrad F, Salazar GA, Richardson JE, Hollingsworth ML, Barraclough

TG, Kelly L, Wilkinson M (2007) A proposal for a standardised protocol to barcode all land plants. *Taxon* 295-299.

Chase MW, Salamin N, Wilkinson M, Dunwell JM, Kesanakurthi RP, Haider N, Savolainen V (2005) Land plants and DNA barcodes: Short-term and long-term goals. *Philosophical Transaction of the Royal Society B Biological Sciences* 360: 1889-1895.

Chilton MD, Drummond MH, Merlo DJ, Sciaky D, MontoyaA, Gordon MP, Nester EW (11977) Stable incorporation of plasmid DNA into higher plant cells: the molecular basis of crown gall tumorigenesis. *Cell* 11(2): 263-271.

Chen J, Zhao J, Erickson DL, Xia N, KressWJ (2015) Testing DNA barcodes in closely related species of *Cucuma* (*Zingiberaceae*) from Myanmar and China. *Molecular Ecology Resources* 15:337-348.

Chen SL, Pang XH, Song JY, Shi LC, Yao H, Han JP, Leon (2014) A renaissance in herbal medicine identification: from morphology to DNA. *Biotechnology Advanced* 32:1237-1244.

Cheng CG, Liu J, Wang H, Xiong W (2010) Infrared spectroscopic studies of Chinese medicines. *Applied Spectroscopy Reviews* 45:165-178.

Chen S, Yao H, Han J, Liu C, Song J, Shi L, Zhu Y, Ma X, Gao T, Pang X, Luo K, Li Y, Li X, Jia X, Lin Y, Leon C (2010) Validation of the *ITS2* region as a novel DNA barcode for identifying medicinal plant species. *PloS One* 5(1): e8613. Doi: 10.1371/journal.pone.0008613.

Chowdhery HJ, Wadhwa BM (1984) Flora of Himachal Pradesh. Vol. 2. BSI, Howrah.

Roy, G.P., Shukla, B.K. & Datt, B. 1992. Flora of Madhya Pradesh. New Delhi.

Cockerham G (1970) Genetical studies on resistance to potato viruses X and Y. *Heredity* 25: 309–348. https://doi.org/10.1038/hdy.1970.35.

Coleman AW (2007) Pan-eukaryote *ITS2* homologies revealed by RNA secondary structure. *Nucleic Acids Research* 35:3322-3329.

Costion C, Ford A, Cross H, Crayn D, Harrington M, Lowe A (2011) Plant DNA barcodes can accurately estimate species richness in poorly known floras. *Plos ONE 6*: e26841.

Cowan MM (1999) Plants products as antimicrobial agents. *Clinical Microbiology Reviews* 654-582.

Crautlein M, Korpelainen H, Pietilainen M, Rikkinen J (2011) DNA barcoding: A tool for improved taxon identification and detection of species diversity. *Biodiversity Conservation* 20:372-389. DOI 10.1007/s10531-010-9964-0.

D'Arcy WD (1979) The classification of the Solanaceae. [In:Hawkes JG,Lester RN and Skelding AD (eds). The Biology and Taxonomy of the Solanaceae. Academic Press: London 3-47.

Daru BH, van der Bank M, Maurin O, Yessoufou K, Schaefer H, Slingsby JA, Davies TJ (2016) A novel phylogenetic regionalization of the phytogeographic zones of southern Africa reveals their hidden evolutionary affinities. *Journal of Biogeography* 43: 155–166.

Deb DB (1980) Enumeration, synonymy and distribution of the Solanaceae in India. *Journal of Economic and Taxonomic Botany* 1:33-54.

Deb DB (1979) Solanaceae in India. In: Hawkes, J.G., Lester, R.N. & Skelding, A.D. The biology and taxonomy of the Solanaceae. 87 -112. London.

Deb DB, Dutta RM (1974) Contribution to the flora ofTirap Frontier Division. J*ournal of the Bombay Natural History Society* 71: 266 -294.

Desalle R, Amato G (2004) the expansion of conservation genetics. *Nature Reviews Genetics* 5:702-712.

Dhellot JR, Matouba MG, Maloumbi JM, Nzikou MG, Dzondo M, Linder M, Desobry S, Parmentier M (2006a) Extraction, chemical composition and nutritional characterization of vegetable oils: case of *Amaranthus hybridus* (var 1 and 2) Longo Brazzaville. *African Journal of Biotechnology* 5:1095-1101.

Doganlar S, Frary A, Daunay MC, Lester RN, Tanksley SD (2002b) A comparative genetic linkage map of eggplant (*Solanum melongena*) and its implications for genome evolution in the Solanaceae. *Genetics* 161: 1697–1711.

Doganlar S, Frary A, Ku HM, Tanksley SD (2002a) Mapping quantitative trait loci in inbred backcross lines of *Lycopersicon pimpinellifolium* (LA1589). *Genome* 45: 1189–1202.

Dudonne S, Vitrac X, Coutiere P, Woillez M, Merillon JM (2009) Comparative study of antioxidant properties and total phenolic content of 30 plant extracts of industrial interest using DPPH, ABTS, FRAP, SOD and ORAC assays. *Journal of Agricultural and Food Chemistry* 57:1768-1774.

Dougnon TV, Bankole HS, Johnson RC, Klotoe JR, Dougnon G, Gbaguidi F, Assogba F, Gbenou J, Sahidou S, Ategbo JM, Rihn BH, Loko F, Boko M, Edorh AP (2012) Phytochemical Screening, Nutritional and Toxicological Analyses of Leaves and Fruits of *Solanum macrocarpon* Linn (Solanaceae) in Cotonou (Benin). *Food and Nutrition Sciences* 3:1595-1603.

Edmonds JM, Chweya JA (1997) Promoting the conservation and use of under-utilized and neglected crops: Black nightshades (*Solanum nigrum* L.) and related species. *International Plants Genetic Resources Institute* 1-90.

Eggleston K, Zhang R, Zeckhauster RJ (2010) The global challenge of antimicrobial resistance: insights from economic analysis. *International Journal of Environmental Research in Public Health* 7(8):3141-3149.

Elekofehinti OO, Kamdem JP, Bolingon AA, Athayde ML, Lopes SR, Waczuk EP, Kade IJ, Adanlawo IG, Rocha JBT (2013) African eggplant (*Solanum anguivi* Lam.) fruit with bioactive polyphenolic compounds exerts in vitro antioxidant properties and inhibits Ca2+- induced mitochondrial swelling. *Asian Pacific Journal of Tropical Biomedicine* 3(10), 757-766. doi:10.1016/S2221-1691(13)60152-5.

Feng S, Jiao K, Zhu Y, Wang H, Jiang M, Wang H (2017) Molecular identification of *Physalis* species (*Solanaceae*) using a candidate DNA barcode: the chloroplast *psbA-trnH* intergenic region. *Genome* 61:15-20.

Fleuret A (1979) The role of wild foidage plants in the diet. A case study from Lushuto, Tanzania. *Ecology and Food Nutrition* 8:89-93.

Friedman J, Yaniv Z, Dafni A, Palewitch D (1986) A preliminary classification of the healing potential of medicinal plants, based on a rational analysis of the ethnopharmacological field survey among Bedouins in the Negev Desert. *Journal of Ethnopharmacology* 16(2):275-287.

Frodin DG (2004) History and concepts of big plant genera. *Taxon* 53:3. https://doi.org/10.2307/4135449.

Galimberti A, Massimo L, Sandionigi A, Bruno A, Mezzasalma V, De Mattia F (2014) DNA barcoding for Minor Crops and Food Traceability. *Advance in Agriculture* 8:831875. http://dx.doi.org/10.11552014/831875.

Ganaie MM, Raja V, Reshi ZA, Verma V (2018) Family Solanaceae: Taxonomy and modern trends. *Annals of Plant Sciences* 7(9):2403-2414.

Gao T, Yao H, Song JY, Zhu YJ, Liu C, Chen SL (2010) Evaluating the feasibility of using candidate DNA barcodes in discriminating species of the large Asteraceae family. *BMC Evolution Biology* 10:324.

Gebhardt C, Ritter E, Barone A et al. (1991) RFLP maps of potato and their alignment with the homoeologous tomato genome. *Theoretical Applied Genetics* 83: 49–57. https://doi.org/10.1007/BF00229225.

Gebhardt C, Ritter E, Debener T et al. (1989) RFLP analysis and linkage mapping in Solanum tuberosum . *Theoretical Applied Genetics* 78: 65–75. https://doi.org/10.1007/BF00299755.

Giovannoni JJ (2007) Fruit ripening mutants yield insights into ripening control. *Current Opinion in Plant Biology* 10(3) 283-289.

Gogoi P, Islam M (2012) Phytochemical screening of *Solanum nigrum* L and *S. myriacanthus* Dunal from District of Upper Assam, India. *IOSR Journal of Pharmacy* 2(3): 455-459.

Gonzalez MA, Baraloto C, Engel J, Mori SA, Petronelli P, Riera B et al. (2009) Identification of Amazonian trees with DNA barcodes. *PlosOne* 4:7483.

Gramazio P, Prohens J, Plazas M (2014) Location of chlorogenic acid biosynthesis pathway and polyphenol oxidase genes in a new interspecific anchored linkage map of eggplant. *BMC Plant Biology* 14: 350.

Grassmann J (2005) Terpenoids as plant antioxidants. *Vitamins and Hormones* 72: 505–535. doi: 10.1016/S0083-6729(05)72015-X.

Guha S, Maheswari SC (1964) In vitro production of embryos from anthers of *Datura*. *Nature* 204: 497.

Gupta AS, Webb RP, Holaday AS, Allen RD (1993) Overexpression of Superoxide Dismutase Protects Plants from Oxidative Stress. *Plant Physiology* 103: 1067-1073.

Hall MB (2007) Methologies challenges in carbohydrate analyses. *Revista Brasileira de Zootecnia* 36 359-367.

Harborne SB, Baxter H (1995) Phytochemical Dictionary. A Handbook of Bioactive Compounds from Plants. Taylor and Francis, London.

Hassan ASH, Bakar AMF (2013) Antioxidative and anticholinesterase activity of *Cyphomandra betacea* fruit. *Science World Journal* 1-7.

He C, Saedler H (2005) Heterotopic expression of MPF2 is the key to the evolution of the Chinese lantern of *Physalis*, a morphological novelty in Solanaceae. *Proceedings of the National Academy of Sciences* 102(16): 5779-5784.

He ZD, Qiao CF, Han QB, Cheng CL, Xu HX, Jiang R, But PPH, Shaw PC (2005) Authentication and quantitative analysis on the chemical profile of cassia bark (cortex cinnamomic) by high-pressure liquid chromatography. *Journal of Agricultural and Food Chemistry* 53: 2423-2428.

Hebert PD, Penton EH, Burns JM, Janzen DH, Hallwachs W (2004) Ten species in one: DNA barcoding reveals cryptic species in the neotropical skipper butterfly *Astraptes fulgerator*. *Proceedings of the National Academy of Sciences USA* 101(4):14812-14817.

Hebert PDN, Stoeckle MY, Zemlak TS, Francis CM (2004b) Identification of birds through DNA barcodes. *PLos Biology* 2: e312.

Hebert PDN, Cywinska A, Ball SL, de Waard JR (2003) Biological identifications through DNA barcodes. *Proceedings of Royal Society London B: Biological Sciences* 270:313-321.

Heinrich M (2015) Quality and safety of herbal medical products: regulation and the need for quality assurance along the value chains. *British Journal of Clinical Pharmacology* 80: 62–66. doi: 10.1111/bcp.12586.

Heinrich M (2007) The identification of medicinal plants. *Journal of Ethnopharmacology* 111:440.

Helenntjaris T, Slocum M, Wright S, Schaefer A, Nienhuis J (1986) Construction of genetic linkage maps in maize and tomato using restriction fragment length polymorphism. *Theoretical Applied Genetics* 72: 761-769.

Heritange J, Zali FHM, Gascoyne-Binzi D, Hawkey PM (1999) Evolution and spread of SHV extended-spectrum beta-lactamases in gram-negative bacteria. *Journal of Antimicrobial Chemotherapy* 44:309-318.

Hirakawa H, Shirawasa K, Miyatake K, Fukuoka H (2014) Draft genome sequence of eggplant (*Solanum melongena* L.): the representative Solanum species indigenous to the old world. *Current Neuropharmacology* 21(6). http://dx.doi.org/10.1093/dnares/dsu027.

Holdsworth WL, Mazourek M (2015) Development of user-friendly markers for the pvr1 and Bs3 disease resistance genes in pepper. *Molecular Breeding* 35: 28.

Hollingsworth PM, Li DZ, Bank M, Twyford AD (2016). Telling plant species apart with DNA: from barcodes to genomes. *Philosophical Transactions of the Royal Society B* 371 : doi: 10.1098/ rstb.2015.0338.

Hollingsworth PM, Graham SW, Little DP (2009) Choosing and using a plant DNA barcode. *Plos One* 6(5): e19254.

Hu P, liang QL, Luo GA, Zhao ZZ, Jiang ZH (2005) Multi-component HPLC fingerprinting of *Radix salvia multiorrhizae* and its LC-MS-MS identification. *Chemical and Pharmaceutical Bulletin* 53: 677-683.

Huang XC, Ci XO, Conran JG, Lie L (2015) Application of DNA barcodes in Asian tropical tree identifictions: how often re ecologists wrong and when does it matter? *Ecology Monographs* 80:267-286.

Ikeno S (1917) Variegation in Plantago. *Genetics* 4(1): 390-416.

Inglis J (1994) Introduction of Ethnobotany. *Nature & Resources* 30:3-4.

Iwu MM (2002) Ethnobotanical approach to pharmaceutical drug discovery: strength and limitations. *Advances in Phytomedicine* 1: 309-320.

Jana S, Shekhawat GS (2010) Phytochemical analysis and antibacterial screening of in vivo and in vitro extracts of Indian medicinal herb: *Anethum graveolens*. *Research Journal of Medicinal Plants* 4: 206-212.

Joshi K, Chavan P, Warude B, Atwardhan B (2004) Molecular markers in herbal drug technology. *Current Science* 87:159-165.

Kala CP, Sajwan BS (2007) Revitalizing Indian Systems of Herbal Medicine by the National Medicinal Plants Board through Institutional Networking and Capacity Building. *Current Science* 93:6.

Kalidass C, Panda PC (2019) The Genus *Solanum* L. (Solanaceae) in Eastern Ghats of India. Regional Plant Resource Centre (RPRC), Bhubaneswar, Odisha. www.rprcbbse.in.

Kalita L, Dash B, Borah U, Deka J, Dash S (2017) Preliminary phytochemical analysis and anti-microbial activity ethanolic extracts of dried fruits of *Solanum torvum* (Family- Solanaceae). *International Journal of Current Pharmaceutical Research* 9(3):123-126.

Kalt W, McDonald JE, Fillmore SA, Tremblay F (2014) Blueberry Effects on Dark Vision and Recovery after Photobleaching: Placebo-Controlled Crossover Studies. *Journal of Agricultural and Food Chemistry* 62: 11180-11189.

Kanungo S, Rout JR, Sahoo SL (2013) Evaluation of Antioxidant Activities in *Withania somnifera* L. in vitro and in vivo Grown Explants. Iran. *Journal of Biotechnology* 11(4), 260-264. Doi:10.5812/ijb.14256.

Kim S, Park M, Yeom SI (2014) Genome sequence of the hot pepper provides insights into the evolution of pungency in *Capsicum* species. *Nature Genetics*46 270-278.

Kim SW, Ban SH, Chung H, Cho S, Chung HJ, Choi PS, Yoo OJ, Liu JR (2004) Taxonomic discrimination of flowering plants by multivariate analysis of Fourier transform infrared spectroscopy data. *Plant Cell Report* 23:246-250. Doi 10.1007/s00299-004-0811-1.

Kitanov G, Karcheva-Bahchevanska D, Lukova P (2015) Comparative analysis of monographs on herbal drugs and herbal drug preparations included in the European Pharmacopoeia (Ph. Eur. 8). *Annales UMCS Pharmacia* 62: 20–27.

Knapp S, Bohs L, Nee M, Spooner DM (2004) Solanaceae – a model for linking genomics with biodiversity. *Comparative and Functional Genomics* 5:285-291.

Kone WM, Kamanzi AK, Terreaux C, Hostettmann K, Traore D, Dosso M (2004) Traditional medicine in North Cote-d'Ivoire screening of 50 medicinal plants for antibacterial activity. *Journal of Ethnopharmacology* 93:43-49.

Kress WJ (2017) Plant DNA barcodes: applications today and in the future. *Journal of Systematic Evolution* 55:291-307.

Kress WJ (2011) DNA barcoding and systematicmonographs. In: Stuessy T, Lack W eds. Monographic plant systematics: Fundamental assessment of plant biodiversity. Berlin: Regnum Vegetable 153. *ARG Gantner Verlag KG* 49-71.

Kress WJ, Erickson DL (2007) A two locus global DNA barcode for land plants: The coding *rbcL* gene complements the non-coding *trnH-psbA* spacer region. *Plos One* 2: e508.

KressWJ, Wurdack KJ, Zimmer EA, Weigt LA, Janzem DH (2005) Use of DNA barcodes to identify flowering plants. *Proceedings of the National Academy of Sciences USA* 102:8369-8374.

Krishnaiah D, Devi T, Bono A, Sarbatly R (2009) Studies on phytochemical constituents of six Malaysian medicinal plants. *Journal of Medicinal Plants Research* 3(2):067-072.

Kumar S, Pandey AK (2014) Medicinal attributes of *Solanum xanthocarpum* fruit consumed by several tribal communities as food: an in vitro antioxidant, anti-cancer and anti-HIV perspective. *BMC Complementary and Alternative Medicine* 14(1):112.

Kumari S, Anokhe A, Kumar R (2017) Nutritional quality and health benefits of Solanaceous vegetables. *Progressive Research – An International Journal* 12:1942-1945.

Lahaye R, Van der Bank M, Bogarin D, Warner J, Pupulin F, Gigot G, Maurin O, Duthoit S, Barraclogh TG, Savolainen V (2008) DNA barcoding the floras of biodiversity hotspots. *Proceedings of the National Academy of Sciences USA* 105: 2923-2928.

Levin RA, Wagner WL, Hoch PC, Nepokroeff M, Pires JC, Zimmer EA, Sytsma KA (2003) Family-level relationships of Onagraceae based on chloroplast *rbcL* and *ndhF* data. *American Journal of Botany* 90: 107–115. doi: 10.3732/ajb.90.1.107 PMID: 21659085.

Li Z, Vickrey TL, McNally MG, Sato SJ, Clemente TE, Mower JP (2019) Assessing Anthocyanin Biosynthesis in Solanaceae as a Model Pathway for Secondary Metabolism. *Genes* 10, 559. Doi:10.3390/genes10080559.

Li XW, Yang Y, Hentry RJ, Rossetto M, Wang Y, Chen S (2015) Plant DNA barcoding: from gene to genome. *Biological Reviews of the Cambridge Philosophical Society* 90:157-166.

Li DZ, Gao LM, Li HT, Wang H, Ge XJ, Liu JQ et al. (2011) Comparative analysis of a large dataset indicates that internal transcribed spacer (*ITS*) should be incorporated into the core barcode for seed plants. *Proceedings of the National Academy of Sciences USA* 108:19641-19646.

Lin H, Zhao JW, Sun L, Chen QS, Zhou F (2011) Freshness measurement of eggs using near infrared (NIR) spectroscopy and multivariate data analysis. *Innovative Food Science & Emerging Technologies* 12(2):182-186.

Lindhout P, Meijer D, Schotte T, Hutten RCB, Visser RGF, Ech HJ (2011) Towards F1 hybrid seed potato breeding. *Potato Research* 54: 301-312.

Lieu PT, Heiskala M, Peterson PA, Yang Y (2001). The roles of iron in health and disease. *Moecular Aspects of Medicine* 22: 1–87. doi: 10.1016/S0098-2997(00) 00006-6.

Liu Z, Zeng X, Yang D, Chu G, Yuan Z, Chen S (2012) Applying DNA barcodes for identification of plant species in the family Araliaceae. *Gene* 499:76-80.

Liu RH (2003) Health benefits of fruits and vegetables are from additive and synergistic combinations of phytochemicals. *The American Journal of Clinical Nutrition* 78:517S-520S.

Lowry O, Rosebrough NJ, Farr AI (1951) Protein measurement with the folin phenol reagent. *Journal of Biological Chemistry* 193: 265-275.

Lu F, Foo LY (1995) Toxicological aspects of food antioxidants. In: Madhavi DL, Desphande SS, Salunkhe DK, editors. Food antioxidants. New York: Marcel Dekker.

Lulekal E, Asfaw Z, Kelbessa E, Van-Damme P (2013) Ethnomedicinal study of plants used for human ailments in Ankober District, North Shewa Zone, Amhara Region, Ethiopia. *Journal of Ethnobiology Ethnomedicine* 9:63.

Luo K, Chen S, Chen K, Song J, Yao H, Ma X, Zhu Y, Pang X, Yu H, Li X, Liu Z (2010) Assessment of candidate plant DNA barcodes using the Rutaceae family. *Science China, Life Sciences* 53(6):701-708.

Malini M, Abirami G, Hemalatha V, Annadurai G (2013) Antimicrobial activity of ethanolic and aqueous extracts of medicinal plants against waste water pathogens. *International Journal of Research in Pure and Applied Microbiology* 3(2):40-42.

Manach C, Scalbert A, Morand C, Remesy C, Jimenez L (2004) Polyphenols: food sources and bioavailability. *The American Journal of Clinical Nutrition* 79: 727–747.

Mann J (1978) Secondary Metabolism. Oxford University press, London, pp. 154.

Manna A, Abalaka ME (2000) Preliminary screening of the various extracts of *Physalis angulata* (L.) for antimicrobial activities. *Spectrum Journal* 7:19-125.

Maobe MAG, Nyarango RM (2013) Fourier Transformer InfraRed Spectrophotometer Analysis of *Warburgia ugandensis* Medicinal herb used for the Treatment of Diabetes, Malaria and Pneumonia in Kisii Region, Southwest Kenya. *Global Journal of Pharmacology* 7(1): 61-68.

Margaryan K, Melyan G, Vardanyan D, Devejyan H, Aroutiounian R (2017) Phenolic content and antioxidant activity of Armenian cultivate and wild grapes. *BIO Web of Conferences* 9:02029. DOI: 10.1051/bioconf/20170902029.

Martin GB, Brommonschenkel S, Chunwongse J et al. (1993a) Map-based cloning of a protein kinase gene conferring disease resistance in tomato. *Science* 262 (5138): 1432-1436.

Matias LJ, Rocha JA, Royo VA, Menezes EV, Junior AFM, de Oliveira DA (2019) Phytochemistry in Medicinal Species of Solanum L. (Solanaceae). *Pharmacognosy Research* 11:47-50.

Mayne ST (2003) Antioxidant nutrients and chronic disease: Use of biomarkers of exposure and oxidative stress status in epidemiologic research. *Journal of Nutrition* 133:933-940.

Mazher M, Malik NZ, Riaz M, Hussain A, Ali Y, Noshad QQ (2016) Phytochemistry and antibacterial assay of fruit leaf and stem extracts of *Solanum nigrum* L. in different solvent. *International Journal of Bioscience* 9(6):129-136.

McCord JM (2001) Analysis of superoxide dismutase activity. *Current Protocol in Toxicology* 23045062.

Mc Donald S, Prenzler PD, Amtolovich M, Robards K (2001) Phenolic content and antioxidant activity of olive extracts. *Food Chemistry* 73:73-84.

Menda N, Semel Y, Peled D, Eshed Y, Zamir D (2004) In silico screening of a saturated mutation library of tomato. *Plant Journal* 38:861-872.

Miao M, Warren A, Song W, Wang S, Shang H, Chen Z (2008) Analysis of the internal transcribed spacer 2 (*ITS2*) region of Scuticociliates and related taxa (*Ciliophora Oligohymenophorea*) to infer their evolution and phylogeny. *Protist* 159:519-533.

Mittler R (2002) Oxidative stress, antioxidants and stress tolerance. *Trends Plant Science* 7: 405-410.

Montoro P, Braca A, Pizza C, De Tommasi N (2005) Structure antioxidant activity relationships of flavonoids isolated from different plant species. *Food Chemistry* 92: 349–355. doi: 10.1016/j.foodchem.2004. 07.028.

Moon JK, Shibamoto T (2009) Antioxidants assays for plants and food components. *Journal of Agricultural and Food Chemistry* 57:1655-1666.

Muellner AN, Schaefar H, Lahaye R (2011) Evaluation of candidate DNA barcoding loci for ecnonomically important timber species of mahogany family (Meliaceae). *Molecular Ecology Resources* 11:450-460.

Mutalib MA, Rahmat A, Ali F, Othman F, Ramasamy R (2017) Nutritional compositions and antiproliferative activities of different solvent fractions from ethanol extract of (tamarillo) fruit. *Malaysian Journal of Medical Sciences* 24: 19-32.

Natural History Museum (NHM) London (2008) About the Family Solanaceae.

Naz S, Siddiqi R, Ahmad S, Rasool SA, Sayeed SA (2007) Antibacterial activity directed isolation of compounds from *Punica granatum*. *Journal of Food Sciences* 72: 341– 345.

Negrutiu I, Jacobs M, Caboche M (1984) Advances in somatic cell genetics of higher plants — the protoplast approach in basic studies on mutagenesis and isolation of biochemical mutants. *Theoretical Applied Genetics* 67: 289–304. https://doi.org/10.1007/BF00272864

Newmaster SG, Fazekas AJ, Steeves RAD, Janovec J (2008) Testing candidate plant barcode regions in the Myristicaceae. *Molecular Ecology Resources* 8:480-490.

Ngan F, Shaw P, But P, Wang J (1999) Molecular authentication of *Panax* species. *Phytochemistry* 50:787-791.

Ningombam DS, Singh PK (2014) Ethnobotanical study of *Phlogacanthus thyrsiformis* Nees: A conserved medicinal plant of Manipur, Northeast India. *International Journal of Herbal Medicines* 1:10-14.

Nithaniyal S, Newmaster SG, Ragupathy S, Krisnamoorthy D, Vassou SL, Parani M (2014) DNA barcode authentication of wood samples of threatened and commercial timber trees within the tropical dry evergreen forest of India. *Plos ONE* 9: e1077669.

Nwankwo IU, Ukaegbu-Obi KM (2014) Preliminary phytochemical screening and antibacterial activity of two nigerian medicinal plants (*Ficus asperifolia and Terminalis catappa*). *Journal of Medicinal Plant and Herbal Therapy Research* 2:1-5.

Nwauzoma AB, Dawari SL (2013) Study on the phytochemical properties and proximate analysis of *Piper umbellatum* (Linn) from Nigeria. *American Journal of Research Communication* 1(7):164-177.

Okpekon T, Yolou S, Gleye C, Roblot F, Loiseau P, Bories C, Grellier F, Frappier F, Laurens A, Hocquemiller R (2004) Antiparasitic activities of medicinal plants used in Ivory Coast. *Journal of Ethnopharmacology* 90:91-97.

Okwu DE (2004) Phytochemicals and Vitamin content of Indigenous Spices of South Eastern Nigeria. *Journal of Sustainable Agricultural Environment* 6:3-34.

Oliveira RCM, Monteiro FS, Silva JLV, Ribeiro LAA, Santos RF, Nascimento RJB, Duarte JC, Agra MF, Silva TMS, Almeida FRC (2006) Methanol and ethyl acetate extracts from *Solanum Megalonyx* Sendtn (Solanaceae) present spasmolytic activity in guinea-pig ileum: A comparative study. *Revista Brasileira de Farmacognosia* 16:146-151.

Olmstead RG, Bohs L (2007) A Summary of molecular systematic research in Solanaceae: 1982-2006. *Acta Horticulture* 745(745):255-268.

Omoti O, Okyi PA (1987) Characterization and composition of the pulp oil and cake of the African pear, *Dacryodes edulis*. *Journal of the Science and Food Agriculture* 38:67-70.

Onyenuga VA, Fetuga (1995) First National seminar on fruits and vegetables. In progress and Recommendation and paper by NITHORT, Ibadan.

Oyeyemi SD, Ayeni MJ, Adebiyi AO, Ademiluyi BO, Tedela PO, Osuji IB (2015) Nutritional Quality and Phytochemical Studies of *Solanum anguivi (*Lam.*)* Fruits. *Journal of Natural Sciences Research* ISSN 2224-3186 (Paper) ISSN 2225-0921 (Online)5: 4.

Pang XH, Song JY, Zhu YJ, Xu H, Huang LF, Chen SL (2011) Applying DNA barcodes for Rosaceae species identification. *Cladistics* 27:165-170.

Panigrahy B, Grumbles LC, Millar D, Naqi SA, Hall CF (1979) Antibiotic induced immunosuppression and Levamisole-induced immunopotentiation in Turkey. *Avian Diseases* 23:401-408.

Peerapattana J, Otsuka K, Hattori Y, Otsuka M (2015). Quantitative analysis of α-mangostin in hydrophilic ointment using near-infrared spectroscopy. *Drug*

Development and Industrial Pharmacy 41: 515–521. doi: 10.3109/03639045.2014.884115.

Podolak I, Galanty A, Sobolewska D (2010) Saponins as cytotoxic agents: a review. *Phytochemistry Reviews*. 9: 425–474. doi: 10.1007/s11101-010-9183-z.

Poojary MM, Vishnumurthy KA, Adhikari AV (2015) Extraction, characterization and biological studies of phytochemicals from *Mammea suriga*. *Journal of Pharmaceutical Analysis* 5: 182–189.

Portis E, Barchi L, Toppino L, Lanteri S, Nazzareno A, Felicioni N, Fusari F, Barbierato V, Cericola F, Vale G, Rotino GL (2014) QTL mapping in eggplant reveals clusters of yield related loci and orthology with tomato genome. *PlosOne*. https://doi.org/10.1371/journal.pone.0089499.

Prince JP, Pochard E, Tanksley SD (1993) Construction of a molecular linkage map of pepper and a comparison of synteny with tomato. *Genome* 36: 404–417.

Quezada NM, Asencio JM, Valle D, Aguilera JM, Gomez B (2006) Antioxidant activity of crude extract, alkaloid fraction and flavonoid fraction from boldo (*Peumusboldus molina*) leaves. *Journal of Food Science* 69: 371–376. doi: 10.1111/j.1365-2621. 2004.tb10700. x.

Ramchiary N, Kehie M, Brahma V (2014) Application of genetics and genomics towards *Capsicum* translational research. *Plant Biotechnology* 8: 101-123.

Ralte L, Bhardwaj U, Singh YT (2021) Traditionally used edible Solanaceae plants of Mizoram, India have high antioxidant and antimicrobial potential for effective phytopharmaceutical and nutraceutical formulations. *Heliyon* 7: e07907.

Ralte L, Vanlalhruaia, Laha RC, Singh YT (2018) Ethnobotanical use of Solanaceae plants of Mizoram, India. *Asian Journal of Microbiology, Biotechnology and Environmental Sciences* 21(1): 229-233.

Rawani A, Ghosh A, Chandra G (2013) Mosquito larvicidal and antimicrobial activity of synthesized nano-crystalline silver particles using leaves and green berry extract of *Solanum nigrum* L. (Solanaceae:Solanales). *Acta Tropica* 128:613-622.

Roberts KM, Daryl KG, Peter AM, Victor WK (2000) Harper's Biochemistry. 25th Edition, Lange Medical Book, Appleton and Lange, pp. 209-210.

Romer P, Hahn S, Jordan T, Strauss T, Bonas U, Lahaye T (2007) Plant pathogen recognition mediated by promoter activation of the pepper Bs3 resistance gene. *Science* 318(5850): 645-648.

Romero CD, Chopin SF, Buck G, Martinez E, Garcia M, Bixby L (2005) Antibacterial properties of common herbal remedies of the southwest. *Journal of Ethnopharmacology* 99:253-257.

Roxburgh W (1795 – 1832) Flora Indica. 3 Vols. Serampore.

Sabran SF, Mohamed M, Abu-Bakar MF (2016) Ethnomedical knowledge of plants used for the treatment of tuberculosis in Johor, Malaysia. *Evidence-based complementary and alternative medicine* eCAM: 2850845.

Sahoo N, Manchikant P (2013) Herbal drug regulation and commercialization: An Indian industry perspective. *Journal of Alternative and Complementary Medicine* 19:957-963.

Sajem AL, Rout J, Nath M (2008) Traditional Tribal knowledge and Status of some Rare and Endemic Medicinal Plants of North Cachar Hills District of Assam, Northeast India. *Ethnobotanical Leaflets* 12:261-275.

Salah N, Miller NJ, Pagangeg G, Tijburg L, Bolwellg P, Rice E, Evans C (1995) Polyphenolic flavonoids s seavengers of aqueous phase radicals as chain breaking antioxidant. *Archives of Biochemistry and Biophysics* 2:339-341.

Saripalli HR, Nandam LS, Teka Z, Prasad M (2013) Preliminary phytochemical studies and efficacy of chloroform extracts of cultured tissues of *Physalis minima* L. against pathogens. *Global Journal of Biology, Agriculture and Health Sciences* 2(4):187-190.

Savolainen V, Cowan RS, Vogler AP, Roderick GK, Lane R (2005) Towards writing the encyclopedia of life: An introduction to DNA barcoding. *Philosophical*

Transactions of the Royal Society London B:Biological Sciences 360:1811-1850.

Sandhar HK, Kumar B, Prasher S, Tiwari P, Salhan M, Sharm P (2011) A review of phytochemistry and pharmacology of flavonoids. *International of Pharmaceuticl Sciences and Research* 1:25-41.

Saxena VK, Sharma RN (1999) Antimicrobial activity of essential oil of *Lantana aculeata*. *Fitoterapia* 70(1):59-60.

Schaneberg BT, Crockett S, Bedir E, Khan IA (2003). The role of chemical fingerprinting: application to Ephedra. *Phytochemistry* 62: 911-918.

Schultz J, Maisel S, Gerlach D, Muller T, Wolf M (2005) A common core of secondary structure of the internal transcribed spacer 2 (*ITS2*) throughout the Eukaryota. *RNA* 11:361-364

Scholthof (2008) Tobacco Mosaic Virus: The beginning of plant pathology. The *American Phytopathological Society* http://dx.doi.org/10.1094/APSnetFeature-2008-0408.

Schuman MC, Balwdwin IT (2016) The layers of plant responses to insect herbivores. *Annual Review of Entomology* 61(1): 373-394.

Schwanninger M, Rodrigues JC, Pereira H, Hinterstoisser B (2004) Effects of short time ball milling on the shape of FT-IR spectra of wood and cellulose. *Vibrational Spectroscopy* 36:23–40.

Shahidi F, Ambigaipalan P (2015) Phenolics and polyphenolics in foods, beverages and spices: antioxidant activity and health effects–A review. *Journal of Functional Foods* 18: 820–897. doi: 10.1016/j.jff.2015.06.018.

Shirawasa K, Isobe S, Hirakawa H, Asamizu E, Fukuoka H, Just D, Rothan C, Sasamoto S, Fujishiro T, Kishida Y (2010) SNP discovery and linkage map construction in cultivated tomato. *DNA Research* 17: 381-391.

Sivastava J, Lambart J, Vietmeyer N (1996) Medicinal plants, an expanding role in development. *Word bank technical paper* 320.

Skoog A, Holler EJ, Crouch SR (2007) Principles of instrumental analysis. 6 Edition, 1039.

Smith BC (2011). Fundamentals of Fourier Transform Infrared Spectroscopy, 2nd Edn. Boca Raton, FL: CRC Press, doi: 10.1201/b10777

Sol Genomics Network (SGN) n.d. About the Solanaceae family. Sol Genomics Network.2008.

Son YO, Kim J, Lim JC, Chung Y, Chung GH (2003) Ripe fruits of *Solanum nigrum* L. inhibits cell growth and induces apoptosis in MCF-7 cells. *Food and Chemical Toxicology* 41(10):1421-1428.

Souza LB, Gindri AL, Fortes TA, Kubica TF, Enderle J, Roehrs R, Silva SM, Manfredini V, Denardin ELG (2020) Phytochemical analysis, antioxidant activity, antimicrobial activity and cytotoxicity of *Chaptalia nutans* leaves. *Advances in Pharmacological and Pharmaceutical Sciences* 326074.

Stankovic N, Mihajilov-Krstev T, Zlatkovic B, Javanovic VS, Mitic V, Jovic J, Comic L, Kocic B, Bernstein N (2016) Antibacterial and antioxidant activity of traditional medicinal plants from Balkan Peninsula. *NJAS-Wageningen Journal of Life Sciences* 78:21-28.

Starr JR, Naczi RF, Chouinard BN (2009) Plant DNA barcodes and species resolution in sedges (*Carex*, Cyperaceae). *Molecular Ecology Resources* 9:151-163.

Stebbins GI (1974) Flowering plants. Evolution Above the species level. Belknap, Cambridge, MA.

Stoeckle M (2003) Taxonomy, DNA and the bar code of life. *Bioscience* 53:2-3.

Stuart B (2004) Infra-red Spectroscopy: Fundamental and Applications, Saunders College Publishing, Philadelphia, PA, USA.

Subashini MS, Rajendran P, Ashok G, Kanthesh B (2015) TLC, FTIR and GC-MS analysis of leaves of *Gymnema sylvestre* R. Br from Kolli Hills, Tamil Nadu, India. *Interntional Journal of Current Microbiology and Applied Sciences* 4(7)757-764.

Sunohara Y, Matsumoto H (2004) Oxidative injury induced by the herbicide quinclorac on EchinochloaoryzicolaVasing and the involvement of antioxidative ability in its highly selective action in grass species. *Plants Science* 167:597-606.

Supiandi MI, Mahanal S, Zubaidah S, Julung H, Ege B (2019) Ethnobotany of traditional medicinal plants used by Dayak desa community in Sintang, West Kalimantan, Indonesia. *Biodiversitas* 20(5): 1264-1270.

Sussex IM (2008) The scientific roots of modern plant biotechnology. *Plant cell* 20(5): 1189-1198.

Taberlet P, Coissac E, Pompanon F, Geilly L, Miquel C (2007). Power and limitations of the chloroplast *trnL* (UAA) intron for plant DNA Barcoding. *Nucleic Acid Research* 35:14-21.

Tamura K, Dudley J, Nei M, Kumar S (2007) MEGA4: molecular evolutionary genetics analysis (MEGA) software version 7.0. *Molecular Biology and Evolution* 24:1596.

Tang L, Kwon SY, Kim SH, Kim JS, Choi JS, Cho KY, Chang KS, Kwak SS, Lee HS (2006) Enhanced tolerance of transgenic potato plants expressing both superoxide dismutase and ascorbate peroxidase in chloroplasts against oxidative stress and high temperature. *Plant Cell Reports* 25: 1380-1386. Doi10.1007/s00299-006-0199-1.

Thomas C (2009) Plant barcode soon to become reality. *Science* 325:526.

Trotter RT, Logan MH (1986) Informant consensus: a new approach for identifying potentially effective medicinal plants. Plant in indigenous medicine and diet, L.N. Etkin, Ed. 91-112, Redgrave Publishing Company, Bedford Hill,NY, USA.

Tsuda T, Horio F, Uchida K, Aoki H, Owasa T (2003) Dietary cyanidin 3-O-β-D-glucoside rich purple corn color prevents obesity and ameliorates hyperglycemia in mice. *Journal of Nutrition* 133: 2125-2130.

Ulewicz-Magulska B, Wesolowski M (2019) Total phenolic contents and antioxidant potential of herbs used for medical and culinary purposes. *Plant Foods for Human Nutrition* 74(1):61-67.

Valko M, Leifritz D, Moncol J, Cronin MTD, Mazur M, Telser J (2007) Free radicals and antioxidants in normal physiological functions and human disease. *Int. J. Biochemistry and Cell Biology* 39:44-84.

Vasco C, Avila J, Ruales J, Svanberg U, Kamal-Eldin A (2009) Physical and chemical characteristics of golden yellow and purple red varieties of tamarillo fruit (*Solanum betaceum* Cav.) *International Journal of Food Sciences and Nutrition* 60: 278-288.

Vasu K, Goud JV, Suryam A, Singara, Chary MA (2009) Biomolecular and phytochemical analyses of three aquatic angiosperms. *African Journal of Microbiology Research* 3(8):418-421.

Venkatappa V (2011) Solanaceae: Flora of Eastern Ghats, (Stylidaceae to Plantaginaceae). Regency Publications, Delhi 4:266-310.

Vijaya PP, Wilson D, Ali AA, Yogananth N, Ali MS, Anuradha V, Parveen PK (2013) Evaluation of in vitro anti-inflammatory and anti-microbial properties of *Pergularia daemia* and *Solanum xanthocarpum. International Journal of Current Microbiology and Applied Science* 2(1):94-99.

Vijayan K, Tsou CH (2010) DNA barcoding in plants: taxonomy in a new perspective. *Current Science* 99:1530-1541.

Vinayaka KS, Nandini KC, Rakshitha MN, Ramya M, Shruthi J, Shruthi VH, Prashith KTR, Raghavendra HL (2010) Proximate composition, Anti-bacterial and Anthelmintic Activity of *Capsicum frutescens* L. var Longa (Solanaceae) Leaves. *Pharmacognosy Journal* 2(12):486-491.

Wang S, Chu Z, Ren M, Jia R, Zhao C, Fei D, Su H, Fan X, Zhang X, Li Y, Wang Y, Ding X (2017) Identification of Anthocyanin Composition and Functional Analysis of an Anthocyanin Activator in *Solanum nigrum* Fruits. *Molecules* 22: 876. Doi:10.3390/molecules22060876.

Wang Y, Yang ZM, Zhang QF, Li JL (2009b) Enhanced chilling tolerance in *Zoysia matrella* by pre-treatment with salicylic acid, calcium chloride, hydrogen peroxide or 6- benzylaminopurine. *Biologia Plantarum* 53: 179-182.

Wattoo JI, Saleem MZ, Shahzad MS, Arif A, Hameed A, Saleem MA (2016) DNA barcoding: amplification and sequence analysis of *rbcL* and *matK* genome regions in three divergent plant species. *Advanced Life Science* 4:3-7.

Weaver CM (2013) Potassium and Health. *Advances in Nutrition* 4: 368S–377S. doi: 10. 3945/an.112.003533.

Wenning M, Seiler H, Scherer S (2002) Fourier transform infrared microspectroscopy, a novel and rapid tool for identification of yeast. *Applied Environmental Microbiology* 68:4717-4721.

Westh H, Zinn CS, Rosdahl VT (2004) An international multicenter study of antimicrobial consumption and resistance in *Staphylococcus aureus* isolates from 15 hospitals in 14 countries. *Microbial Drug Resistance* 10:169-176.

Wettstein WR (1981) Solanaceae. *Nat Planzenfam* 4(3B):4-38.

Wilf P, Carvalho MR, Gandolfo MA, Ruben N (2017) Eocene lantern fruits from Gondwanan Patagonia and the early origins of Solanaceae. *Science* 355(6320):71-75.

Xiao LQ, Moller M, Zhu H (2010) High nrDNA *ITS* polymorphism in the ancient extant seed plant Cycas: incomplete concerted evolution and the origin of pseudogenes. *Molecular Phylogenetics and Evolution* 55:168-177.

Xin T, Yao H, Gao H, Zhou X, Xiaochong M, Xu C, Chen J, Han J, Pang X, Xua R, Song J and Ngan F, Shaw P, But P, Wang J (1999) Molecular authentication of *Panax* species. *Phytochemistry* 50:787-791.

Xue CY, Li DZ (2011) Use of DNA barcode *Sensu lato* to identify traditional Tibetan medicinal plant *Gentianopsis paludosa* (Gentianaceae) *Journal of Systematic Evolution* 49(3): 267-270.

Yadav VK (2018) Phytochemical and pharmacological studies of *Blumea lacera. International Journal of Green Pharmacy* 12(1).

Yan-Hwa C, Chao-Lin C, Hsia-Fn H (2000). Flavonoid content of several vegetables and their antioxidant activity. *Journal of the Science of Food and Agriculture* 80: 561-566.

Yang IC, Tsai CY, Hsieh KW, Yang CW, Ouyang F, Lo Y M et al. (2013). Integration of SIMCA and near-infrared spectroscopy for rapid and precise identification of herbal medicines. *Journal of Food and Drug Analysis* 21: 268–278. doi: 10.1016/j.jfda.2013.07.008.

Yao H, Song J, Liu C, Luo K, Han J, Li Y, Pang X, Xu H, Zhu Y, Xiao P, Chen S (2010) Use of *ITS2* region as the universal DNA barcode for plants and animals. *Plos One* 5: e13102.

Yessoufou K, Davies JT, Maurin O, Kuzmina M, Schaefer H, Van der Bank M, Savolainen V (2013) Large herbivores favour species diversity but have mixed impacts on phylogenetic community structure in an African savanna ecosystem. *Ecology* 101: 614–625.

Yeung AWK, Heinrich M, Kijjoa A, Tzvetkov NT, Atanasov AG (2020) The ethnopharmacological literature: An analysis of the scientific landscape. *Journal of Ethnopharmacolgy* 250:112414.

Yin JX, Deng XH, Che XY, Zhang LH (2005) Study on TLC identification of *Fructus xanthii. WCJPS.* 20: 067-069.

Yousaf Z, Wang Y, Baydoun E (2013) Phytochemistry and Pharmacological Studies on *Solanum torvum* Swartz. *Journal of Applied Pharmaceutical Science* 3(04): 152-160.

Yu Q, Osborne L, Rengel Z (1998) Micronutrient deficiency changes activities of superoxide dismutase and ascorbate peroxidase in tobacco plants. *Journal of Plant Nutrition* 2(7): 1427-1437. doi:10.1080/01904169809365493.

Zhang L, Tu ZC, Xie X, Lu Y, Wang ZX, Wang H (2016) Antihyperglycemic, antioxidant activities of two *Acer palmatum* cultivars, and identification of phenolics profile by UPLC-QTOF-MS/MS: new natural sources of functional constituents. *Industrial Crops and Products* 89: 522–532. doi: 10.1016/j.indcrop.2016.06.005.

Zhang YB, Shaw PC, Sze CW, Wang ZT (2007) Molecular authentication of Chinese herbal materials. *Journal of Food and Drug Analysis* 15:1-9.

Zhang R, Eggleston K, Rotimi V, Zeckhauser RJ (2006) Antibiotic resistance as a global threat: evidence from China, Kuwait and the United States. *Global Health* 7(2):6.

Zhang H, Xu X, Zhang J (2003) Nutrition ingredient and exploitation of *Solanum nigrum* L. *China Wild Plant Research* 23: 44-46.

Zhang JH, Wang ZW, Zhang X, Jin JQ, Yang FL (2000) Authentication of *Tribulus terrestris* L. by morphology, microscope and TLC. *Zhong Yao Cai* 23:675-677.

Zhao L, Huang C, Shan Z, Xiang B, Mei L (2005) Fingerprint analysis of *Psoralea corifolia* L. by HPLC and LC-MS. *Journl of Chromatography B: Analytical in the Technologies in the Biomedical and Life Sciences* 821: 67-74.

Zheng W, Wang SY (2001) Antioxidant activity and phenolic compounds in selected herbs. *Jorunal of Agricultural and Food Chemistry* 49:5165-5170.

CPSIA information can be obtained
at www.ICGtesting.com
Printed in the USA
BVHW091958200123
656721BV00007B/794

9 781044 470004